Membrane Proteins

A Laboratory Manual

Edited by
A. Azzi U. Brodbeck P. Zahler

With Contributions by

P. R. Allegrini A. Azzi N. A. Bersinger K. Bill O. J. Bjerrum
A. Boschetti G. Brandolin U. Brodbeck C. Broger J. Brunner
Z. I. Cabantchik R. P. Casey K. J. Clemetson J. Doussiere
W. Eichenberger B. W. Fulpius R. Gentinetta R. W. James
C. Kempf G. J. M. Lauquin H. Lüdi M. Müller H. Oetliker P. Ott
F. Pattus V. Riess B. Roelofsen C. R. Römer-Lüthi C. Rothen
D. Sargent E. Sauton-Heiniger P. Schoch B. Schwendimann
H. Sigrist K. Sigrist-Nelson M. Thelen P. V. Vignais H. Weiss
P. Wingfield M. Wolf C. Wüthrich P. Zahler M.-L. Zahno

With 76 Figures

Springer-Verlag
Berlin Heidelberg New York 1981

Professor Dr. ANGELO AZZI
Professor Dr. URS BRODBECK
Medizinisch-Chemisches Institut
Universität Bern
Bühlstrasse 28
CH-3012 Bern

Professor Dr. PETER ZAHLER
Biochemisches Institut
Universität Bern
Freiestrasse 3
CH-3012 Bern

ISBN 3-540-10749-5 Springer-Verlag Berlin Heidelberg New York
ISBN 0-387-10749-5 Springer-Verlag New York Heidelberg Berlin

Offsetprinting and bookbinding: Beltz, Offsetdruck, Hemsbach/Bergstr.
2131/3130-543210

Preface

The growing interest in the field of biological membranes in recent years is documented by the very large number of articles, reviews, journals and books which are appearing in this field.

Why then now a manual on "Membrane Proteins"?

The answer is multifold. The protocols which were distributes by the teachers and lecturers at the FEBS-SKMB Course organized in Bern appeared to be very useful not only during the Course to correctly perform the experiments, but also for the future benefit of other students and other curses. To us they appeared very modern and of simple execution, ideal for a University Advanced Course, a Summer School, or similar scientific initiatives. The possibility was also foreseen that such a manual could be used by professional scientists, although not initiated into the problems, assumptions and intricacies of biochemical methodology.

There are also many research teams who study proteins, for example of human fluids, and who will certainly be interested in the application of new but simply described methods. At the same time we present the student with some more complicated physical techniques which are, however, simply described and easy to execute.

The techniques described in the manual have been improved by the feedback reactions of the students who participated in the Course, and have been updated during the summer of 1980 to make them as modern as possible. Some important experiments which were not performed at the Course have been also added, in order to give more completeness to the manual.

This manual follows that edited by E. Carafoli and G. Semenza entitled *Membrane Biochemistry,* published as a result of a few Advanced Courses in Bioenergetics. Of this, needless to say, our initiative represents a logical continuation and a completion. Finally our thanks to the authors, who made an essential contribution in the writing and adapting the experimental protocols, and to the students, who provided the necessary feed-back. Without either of them this manual would not have been realized.

Bern, June 1981
ANGELO AZZI
URS BRODBECK
PETER ZAHLER

Contents

Reconstitution

Modification

Characterization — Spectral Techniques

Characterization — Other Techniques

Contributors

Allegrini, P.R., Biochemisches Institut, Universität Bern, Bern, Schweiz

Azzi, A., Medizinisch-chemisches Institut, Universität Bern, Bern, Schweiz

Bersinger, N.A.,Département de Biochimie Science II, Université de Genève, Genève, Suisse

Bill, K., Medizinisch-chemisches Institut, Universität Bern, Bern, Schweiz

Bjerrum, O.J., The Protein Laboratory, University of Copenhagen, 34 Sigurds-gade, Copenhagen, Denmark

Boschetti, A., Biochemisches Institut, Universität Bern, Bern, Schweiz

Brandolin, G., Laboratoire de Biochimie (INSERM U.191 et CNSR-ERA no 903), Départment de Recherche Fondamentale, Centre d'Etudes Nucléaires, 85X, Grenoble, France

Brodbeck, U., Medizinisch-chemisches Institut, Universität Bern, Bern, Schweiz

Broger, C., Medizinisch-chemisches Institut, Universität Bern, Bern, Schweiz

Brunner, J., Laboratorium für Biochemie, Eidgenössische Technische Hoch-schule (ETH), Zürich, Schweiz

Cabantchik, Z.I., Department of Biological Chemistry, Institute of Life Sciences, The Hebrew University of Jerusalem, Jerusalem, Israel

Casey, R.P., Medizinisch-chemisches Institut, Universität Bern, Bern, Schweiz

Clemetson, K.J., Theodor Kocher Institut, Universität Bern, Bern, Schweiz

Doussiere, J., Laboratoire de Biochimie (INSERM U.191 et CNRS-ERA no 903), Département de Recherche Fondamentale, Centre d'Etudes Nucléaires, 85X, Grenoble, France

Eichenberger, W., Biochemisches Institut, Universität Bern, Bern, Schweiz

Fulpius, B.W. Département de Biochimie Science II, Université de Genève, Genève, Suisse

Gentinetta, R., Medizinisch-chemisches Institut, Universität Bern, Bern, Schweiz

James, R.W.,Département de Biochimie Science II, Université de Genève, Genève, Suisse

Kempf, C., Biochemisches Institut, Universität Bern, Bern, Schweiz

Lauquin, G.J.M., Laboratoire de Biochimie (INSERM U.191 et CNRS-ERA no 903), Département de Recherche Fondamentale, Centre d'Etudes Nucléaires, 85X, Grenoble, France

Lüdi, H., Medizinisch-chemisches Institut, Universität Bern, Bern, Schweiz

Müller, M., Laboratorium für Elektronenmikroskopie, Eidgenössische Technische Hochschule I (ETH), Zürich, Schweiz

Oetliker, H., Physiologisches Institut, Universität Bern, Bern, Schweiz

Ott, P., Medizinisch-chemisches Institut, Universität Bern, Bern, Schweiz

Pattus, F., Centre de Biochimie et Biologie Moléculaire, 31 Chemin Joseph Aiguier, Marseille, France

Riess, V., Medizinisch-chemisches Institut, Universität Bern, Bern, Schweiz

Roelofsen, B., Laboratory of Biochemistry, University of Utrecht, Padualaan 8, Utrecht, The Netherlands

Römer-Lüthi, C.R., Medizinisch-chemisches Institut, Universität Bern, Bern, Schweiz

Rothen, C., Theodor Kocher Institut, Universität Bern, Bern, Schweiz

Sargent, D., Institut für Molekularbiologie und Biophysik, Eidgenössische Technische Hochschule (ETH), Zürich, Schweiz

Sauton-Heiniger, E., Biochemisches Institut, Universität Bern, Bern, Schweiz

Schoch, P., Institut für Molekularbiologie und Biophysik, Eidgenössische Technische Hochschule (ETH), Zürich, Schweiz

Schwendimann, B., Département de Biochimie Science II, Université de Genève, Genève, Suisse

Sigrist, H., Biochemisches Institut, Universität Bern, Bern, Schweiz

Sigrist-Nelson, K., Medizinisch-chemisches Institut, Universität Bern, Bern, Schweiz

Thelen, M., Medizinisch-chemisches Institut, Universität Bern, Bern, Schweiz

Vignais, P.V., Laboratoire de Biochimie (INSERM U.191 et CNRS-ERA no 903), Département de Recherche Fondamentale, Centre d'Etudes Nucléaires, 85X, Grenoble, France

Weiss, H., Europäisches Laboratorium für Molekularbiologie, Universität Heidelberg, Heidelberg, Bundesrepublik Deutschland

Wingfield, P., Europäisches Laboratorium für Molekularbiologie, Universität Heidelberg, Heidelberg, Bundesrepublik Deutschland

Wolf, M., Biochemisches Institut, Universität Bern, Bern, Schweiz

Wüthrich, C., ISREC, Université de Lausanne, Lausanne, Suisse

Zahler, P., Biochemisches Institut, Universität Bern, Bern, Schweiz

Zahno, M.-L., Theodor Kocher Institut, Universität Bern, Bern, Schweiz

Analytical Techniques

Two-Dimensional Electrophoresis of Membrane Proteins

A. BOSCHETTI, E. SAUTON-HEINIGER, and K.J. CLEMETSON

I. Introduction

If protein mixtures containing more than approximately 30 components have to be analyzed, one-dimensional electrophoretic separation methods (on disc-, SDS-, and/or gradient gels) often give insufficient resolution of the individual protein bands. In such cases a two-dimensional electrophoretic separation would be helpful, which uses in each dimension a different physicochemical parameter for polypeptide separation, such as electrical charge and molecular weight.

Supramolecular, biological structures, e.g., membranes or ribosomes, contain far more than 30 individual protein components. Indeed, research on the topology of the ribosome began with the development of a two-dimensional separation method for the individual ribosomal proteins by Kaltschmidt and Wittmann (1970). They separated the ribosomal proteins in the first dimension according to charge differences using about neutral pH and low acrylamide concentration, and in the second dimension according to molecular weight in a gel with acid pH and high acrylamide concentration. This method, however, is not applicable to membrane proteins due to their insolubility in normal buffers. For the solubilization of membrane proteins detergents and chaotropic reagents have to be used, which dissociate them into monomeric units (Ames and Nikaido 1976; Bhadi et al. 1974; Boschetti et al. 1978a; Boschetti et al. 1978b; Cabral et al. 1976; Clemetson et al. 1979; Novak-Hofer and Siegenthaler 1977; O'Farrell 1975; Piperno et al. 1977; Rubin and Milikowski 1978).

With the method used here membrane proteins are separated in the first dimension by isoelectric focusing in Triton-X-100/urea according to their isoelectric points. In this medium most but not all proteins are solubilized very efficiently. In the second dimension proteins are separated according to their molecular weight using SDS-gel-electrophoresis according to Laemmli (1970) but with an acrylamide gradient-gel to enhance resolution (Boschetti et al. 1978b).

In this experiment the proteins of five different membranes will be separated simultaneously.

II. Solutions, Equipment, Starting Material

A. Stock Solutions

Acrylamide Solution
300 g acrylamide (Merck, zur Synthese)
  8 g N,N'-methylenediacrylamide (Merck, zur Synthese) recryst.
made up to 1000 ml with H_2O

Gel Buffer (Separating Gel Buffer of Laemmli)	final conc.
181.7 g Tris-Base (Merck, LAB)	1.5 M
4 g SDS (Serva)	0.4 %

made up to 800 ml with H_2O
brought to pH 8.8 with conc. HCl (Merck p.a.)
made up to 1000 ml with H_2O

Marker Proteins	Mol. weight
β-galactosidase (Worthington)	116 000
phosphorylase a, rabbit (Sigma)	92 500
transferrin, human (Miles)	77 000
bovine serum albumin, cryst. (Sigma)	67 000
glutamate dehydrogenase (Böhringer)	53 000
ovalbumin, cryst. (Serva)	45 000
chymotrypsinogen, cryst. (Serva)	25 000
lysozmye, grade I (Sigma)	14 300

Each protein is stored in small aliquots at 1 mg/ml in H_2O either at 4°C or frozen. Before use, equal volumes of each protein solution are mixed together with 10% SDS-solution to give a final concentration of 0.5% SDS. The mixture is heated 5 min at 90°C. 1/10 volume of a solution of 60% sucrose, 0.2% bromophenol blue, 10 mM Tris/HCl pH 7.6 is added.

B. Reagents for Solubilization of the Samples

	Final conc.	
urea, reinst (Merck)	6	M
Triton-X-100 (NEN)	2.5	%
Servalyte 2-11, 40% (Serva)	2	%
β-mercaptoethanol (Serva)	0.4	M

C. For Electrofocusing (1st Dimension)

Gel solution (for 12 rod gels, 6 mm diameter, 9.5 cm long)	Final conc.	
23.4 g urea (Merck, reinst)	8	M
2.25 g sucrose (Merck, biochem. Zwecke)	5	%
6.72 ml acrylamide stock solution	4	%
1.12 ml 3.5% N,N′-methylene diacrylamide (Merck) recryst.		
in H_2O	0.2	%
1.95 ml 25% Triton-X-100 (NEN)	1.1	%
2.43 ml 40% Servalyte pH 2-11 (Serva)	2	%
made up to 43.8 ml with H_2O		

Just before filling the glass tubes add:

30 µl N,N,N′,N′-tetramethylethylene diamine	0.066%
1.2 ml ammonium persulphate solution (15 mg/ml)	
freshly prepared	0.04 %

Electrolyte Solutions
1000 ml   2% H_2SO_4 for lower, anodic compartment
1000 ml   0.2 M NaOH for upper, cathodic compartment

D. Measurement of pH-Gradient in the Gel

Double-distilled water, boiled for 5 min and kept under nitrogen gas.

E. For Dialysis of 1st Dimension Gels

	Final conc.	
5 ml gel buffer stock solution	30	mM
90 g urea (Merck, reinst)	6	M
12.5 ml 10% SDS (Serva)	0.5	%
made up to 250 ml with H_2O		

F. For SDS-Gel Electrophoresis (2nd Dimension)

Amount for 5 slab gels (145 × 115 × 2.5 mm)

Separating Gel

	Heavy solution		Light solution	
	ml	Final conc.	ml	Final conc.
acrylamide stock solution	66.6	20%	26.6	8%
gel buffer stock solution	25	0.375 M	25	0.375 M
glycerol 87% (Merck p.a.)	to 100	ca. 7.5%	–	–
H_2O	–	–	to 100	–

Just before use add	To *17 ml* heavy solution		To *16 ml* light solution	
N,N,N′,N′-tetramethyl-ethylene diamine	10 μl	0.06%	10 μl	0.06%
ammonium persulphate solution (16 mg/ml) *freshly prepared*	0.3 ml	0.026%	0.4 ml	0.037%

Upper Gel

			Final conc.	
10	ml	acrylamide stock solution	3	%
2	ml	3.5% N,N′-methylene-diacrylamide (Merck), recryst. in H_2O	0.7	%
2	ml	10% SDS (Serva)	0.2	%
25	ml	gel buffer stock solution	0.375	M
6	g	urea	1	M
to 98.6 ml with H_2O				

Just before use add:

1	ml	N,N,N′,N′-tetramethyl-ethylene-diamine	1	%
0.4 ml		ammonium persulphate solution (*150* mg/ml) *freshly prepared*	0.06	%

Running Buffer

		Final conc.	
15.15 g	Tris base (Merck, LAB)	0.025	M
72 g	glycine (Merck, for medical purp.)	0.192	M
10 g	SDS (Serva)	0.2	%
made up to 5000 ml with H_2O			

G. Staining Solution   Final conc.

25 g	Coomassie brilliant blue 250	0.5 %
350 ml	acetic acid conc.	7 %
1250 ml	ethanol	25 %

made up to 5000 ml with H_2O; filtered

H. Destaining Solution

a) 350 ml acetic acid conc.   7 %
    1250 ml ethanol   25 %
    made up to 5000 ml with H_2O

b)   7 %   acetic acid in H_2O (for gel storage)

I. Special Equipment

Most convenient is the apparatus for two-dimensional gel electrophoresis (Boschetti et al. 1978a) modified from that originally described by Kalt-schmidt and Wittmann (1970). Other equipment for electrophoresis, such as described by Clemetson et al. (1979) or Studier (1973) can be used as well

power supply, 200 mA, 500 V

pH-glass electrode for measurement on the surface of gels: electrode from Ingold AG, Urdorf, Switzerland, stand from Desaga, Heidelberg.

J. Starting Material

1. Thylakoid membranes from the unicellular green alga *Chlamydomonas reinhardi,* isolated as 5–30 000 *g* sediment of the cell homogenate without further purification (Boschetti et al. 1978a). 1 mg/ml Chlorophyll in 10 mM Tris/HCl pH 7.6 + 2 mM $MgCl_2$.

2. Thylakoid membranes from isolated spinach chloroplasts prepared according to Boschetti et al. (1978a). 1 mg/ml Chlorophyll in 10 mM Tris/HCl ph 7.6 + 2 mM $MgCl_2$.

3. Ghosts from human erythrocytes prepared according to Dodge et al. (1963). The initial concentration of 15 mg/ml protein in 10 mM Tris/HCl pH 7.6 was lowered by dilution with buffer to 6 mg/ml.

4. Membranes from human platelets prepared according to Käser-Glanzmann et al. (1977). The suspension with approx. 10 mg/ml protein in 0.15 M NaCl, 0.01 M Tris/HCl pH 7.4 was diluted 1:1 with 0.01 M Tris/HCl pH 7.6.

5. Membranes from P-815 cells (mouse mastocytoma, i.e., mast cell tumor) prepared according to Clemetson et al. (1976). The suspension with approx. 10 mg/ml protein in 0.15 M NaCl, 0.005 M Tris/HCl pH 7.4 was diluted 1:1 with 0.01 M Tris/HCl pH 7.6.

III. Experimental Procedures

The experiment can be divided into five steps:

A. Preparation of gels
B. Solubilization of samples
C. Electrofocusing (1st dimension)
D. SDS-electrophoresis (2nd dimension)
E. Staining of the gels

If during a course, the experiment must be carried out in one day, the five above-mentioned steps cannot be done in chronological order. Therefore one group performs in the morning the second-dimensional separation of the experiment of the previous group and prepares in the afternoon the first dimension separation for the following group. Hence, the order of steps will be D, A, B, C, E.

A. Preparation of the Gels

1. Slab Gels for 2nd Dimension

The apparatus of Kaltschmidt and Wittmann (1970) has been modified so that the gel chambers are 12 cm high, 14.5 cm wide and 0.25 cm thick. The Plexiglas plates on both sides of the gel-slabs are cut so that a V-shaped trough is formed at the top, in which the first-dimensional gel rods can be laid.

The six parts of the apparatus are assembled and the bottom of the chambers sealed with adhesive tape. The chambers thus formed are filled first with 1 cm of water, then from the bottom to the height of 11.5 cm with the linear polyacrylamide gradient described in 1.6 as separation gel. (Don't forget to add first the TEMED and ammonium persulfate!) Polymerization occurs within 2 h at room temperature.

2. Rod Gels for 1st Dimension

Twelve glass tubes (12 cm length, 0.6 cm inner diameter) are placed into the "polymerization rack". After adding TEMED and ammoniumpersulfate to the gel solution described in II,C,1, the tubes are filled to a height of 9.5 cm by means of a syringe. Immediately after that, 0.5 cm of water is overlayered. Close the tubes with Parafilm until they are used.

B. Solubilization of the Samples

The starting material listed in II, I is solubilized as follows: 0.5 ml of the samples 1, 2 and 3 in polypropylene tubes are supplemented with 55 μl 10% SDS and heated for 5 min at 60°C. After cooling to each are added:

356 mg	urea
55 μl	Servalyt (40%)
100 μl	Triton-X-100 (25%)
22 μl	β-mercaptoethanol

The mixture is kept at 40°C for 30 min with agitation every 10 min. The samples are transferred with a Pasteur pipette to centrifuge tubes and centrifuged for 30 min at 55,000 g in the Type 50 rotor at 20°C (24,000 rpm). The supernatants are ready to electrofocus.

To 0.1 ml of samples 4 and 5 in polypropylene tubes are added 11μl of 10% SDS. After 5 min at 90°C and cooling to each are added:

70 mg	urea
11 μl	Servalyt (40%)
20 μl	Triton-X-100 (25%)
5 μl	β-mercaptoethanol

The mixture is kept at 40°C for 30 min. After cooling the suspension is ready to electrofocus.

C. Electrofocusing (1st Dimension)

When the rod gels have polymerized, i.e., when the interface between the gel and the overlayered water becomes visible as a sharp line, the water on top is completely removed with the aid of paper tissues.

After the glass tubes have been inserted into the electrophoresis chamber, the bottoms of the tubes are covered with plastic caps with slits 2 to 3 mm wide, which prevent the gels from falling out during electrofocusing.

In adding the 2% sulfuric acid to the lower electrode compartment, care should be taken that no air bubbles are trapped in the ends of the tubes. Now the samples are loaded with a microsyringe on top of the gels according to the following scheme. (Save rest of the samples as reference for second dimensional electrophoresis.)

No. of tubes	1	2	3	4	5	6	7	8	9	10	11
Sample	1	1	1	2	2	2	3	3	3	4	5
μg Protein	400	400	200	400	400	200	700	700	350	400	400
Approx. μl	200	200	100	200	200	100	200	200	100	100	100
Used for	2 D	pH	st.	2 D	pH	st.	2 D	pH	st.	2 D	2 D

2 D = Second dimension; st. = stain

The samples are carefully overlayered with 0.2 M NaOH and then the upper electrode compartment is filled with the same solution. Electrofocusing is carried out at room temperature overnight at 260 V.

D. SDS-Electrophoresis (2nd Dimension)

1. Dialysis of the 1st-Dimensional Gels

After electrofocusing overnight the gel rods No. 1, 4, 7, 10, and 11, which will be used for the second-dimensional electrophoresis, are removed from the glass tubes using a syringe with a fine needle. The gels are dialyzed at room temperature in test tubes by leaving them for 6 × 15 min in the dialysis buffer described in 1.5.

2. 2nd-Dimensional Electrophoresis

The adhesive tapes are taken off the electrophoresis apparatus. The water on top of the slab gels is removed with a syringe and the dialyzed gel rods are laid on top of the slab gels. In order to form wells for coelectrophoresis of reference and marker proteins, approx. 5-mm-wide pieces of wood are introduced on both sides of each rod gel between the Plexiglas plates in the upper gel (which will be applied later). Now the upper gel solution is added rapidly on top of the slab gel, so that the rod gel is completely embedded (polymerization occurs almost immediately!). The electrode chambers are filled with running buffer (see II, F, 3). The pieces of wood are removed. To the bottom of one well 50 μl of a solution of marker proteins are added with a syringe. To the other well is added approx. 20–30 μl (70 μg protein) of solubilized sample (saved in III, C) which has been supplemented with 10% SDS to a final concentration of 0.5% SDS and heated 5 min at 60°C.

With the anode at the bottom, electrophoresis is performed initially at 180 mA constant current. During electrophoresis the current is reduced gradually, so that never more than 17 Watts per 5 plates are applied. After 6–7 h, when the β-front has reached the bottom of the slab gels, electrophoresis is stopped.

3. pH-Measurement in the Gels

The pH-gradient in the rod gels after electrofocusing can be determined either by elution or directly with a special electrode. For these measurements the gel rods No. 2, 5, and 8 can be used.

The gels are removed from the glass tubes with a syringe and laid onto a V-shaped support.

a) pH-Measurement by Elution. The gel is cut with a scalpel into pieces, 0.5 cm long, each of which is kept in a covered tube in 1 ml of CO_2-free water under N_2 (II, D). After at least 4 h standing at room temperature the pH is measured with a combined glass electrode.

b) pH-Measurement with Special Electrode. The equipment specified in paragraph II, I is used. Each 0.5 cm the pH is measured by pressing the tips of the electrode onto the surface of the gel. Care should be taken that both tips make contact.

The pH-gradients measured by both methods are drawn on the same graph.

E. Staining of the Gels

After completion of the 2nd-dimensional electrophoresis the apparatus is disassembled and the slab gels carefully transferred onto the staining grids (upper gel can be cut off). They are fixed overnight in 7% acetic acid, then stained 2–4 h in the staining solution (see II, G) and afterwards destained (see II, H) until the background shows only a faint blue color. The final destaining occurs during storage of the gels in 7% acetic acid. The gels can be photographed on a light box with Ilford Pan F film and orange filter.

If first-dimensional gel rods are to be stained, the gels No. 3, 6 and 9 are removed from the glass tubes, fixed for at least 6 × 15 min in 7% acetic acid (to remove the ampholyte) and stained and destained in test tubes at room temperature by the same method as the slab gels.

References

Ames GFL, Nikaido K (1976) Two-dimensional gel electrophoresis of membrane proteins. Biochemistry 15:616–623

Bhakdi S, Knüfermann H, Wallach DFH (1974) Separation of EDTA-extractable erythrocyte membrane proteins by isoelectric focusing linked to electrophoresis in sodium dodecyl sulfate. Biochim Biophys Acta 345:448–457

Boschetti A, Sauton-Heiniger E, Schaffner JC, Eichenberger W (1978a) A two-dimensional separation of proteins from *Chlamydomonas* thylakoids and other membranes. Physiol Plant 44:134–140

Boschetti A, Diezi R, Eichenberger W, Schaffner JC (1978b) Characterization of thylakoid proteins by two-dimensional separation. In: Akoyunouglou G et al. (eds) Chloroplast development. Elsevier/North-Holland Biomedical Press, Amsterdam New York, pp 195–200

Cabral F, Saltzgeber J, Birchmeier W, Deters D, Frey T, Kohler C, Schatz G (1976) Structure and biosynthesis of cytochrome c oxidase. In: Bücher T, Neupert W, Sebald W, Werner S (eds) Genetics and biogenesis of chloroplasts and mitochondria. North Holland Publ Co, Amsterdam, pp 215–230

Clemetson KJ, Gerber A, Bertschmann M, Lüscher EF (1976) Solubilization of histocompatibility and tumor-associated antigens of the P-815 murine mastocytoma cell. Eur J Cancer 12:263–270

Clemetson KJ, Capitano A, Lüscher EF (1979) High resolution two-dimensional gel electrophoresis of the proteins and glycoproteins of human blood platelets and platelet membranes. Biochim Biophys Acta 553:11–24

Dodge JT, Mitchell C, Hanahan DJ (1963) The preparation and chemical characteristics of hemoglobin-free ghosts of human erythrocytes. Arch Biochem Biophys 100:119–130

Kaltschmidt E, Wittmann HG (1970) Ribosomal proteins. VII. Two-dimensional polyacrylamide gel electrophoresis for fingerprinting of ribosomal proteins. Proc Natl Acad Sci USA 67:1276–1282

Käser-Glanzmann R, Jakábová M, George JN, Lüscher EF (1977) Stimulation of Calcium uptake in platelet membrane vesicles by adenosine 3',5'-cyclic monophosphate and protein kinase. Biochim Biophys Acta 466:429–440

Laemmli WK (1970) Cleavage of structural proteins during the assembly of the head of bacteriophage T4. Nature (London) 227:680–685

Novak-Hofer I, Siegenthaler PA (1977) Two-dimensional separation of chloroplast membrane proteins by isoelectric focusing and electrophoresis in sodium dodecyl sulfate. Biochim Biophys Acta 468:461–471

O'Farrell PH (1975) High resolution two-dimensional electrophoresis of proteins. J Biol Chem 250:4007–4021

Piperno G, Huang B, Luck DJL (1977) Two-dimensional analysis of flagellar proteins from wild-type and paralyzed mutants of *Chlamydomonas reinhardi*. Proc Natl Acad Sci USA 74:1600–1604

Rubin RW, Milikowski C (1978) Over two hundred polypeptides resolved from the human erythrocyte membrane. Biochim Biophys Acta 509:100–110

Studier FW (1973) Analysis of bacteriophage T7 early RNas and proteins on slab gels. J Mol Biol 79:237–248

Analysis of Membrane Antigens by Means of Quantitative Detergent-Immunoelectrophoresis

O. J. BJERRUM

I. Introduction and Aims

Most methods available for the characterization of a given protein require the protein to be available in pure form and in relatively large amounts. These requirements are difficult to fulfil for most membrane proteins because of the problems involved in their isolation. We have therefore taken an interest in some methods that are not limited by these requirements. In contrast to most other immunological methods, quantitative immuno-electrophoretic methods are sensitive, simple to perform, and permit the distinction and characterization of a multitude of different antigens at the same time. It is thus unnecessary to purify individual membrane proteins or to produce specific antibodies, since crude protein mixtures can be examined directly using polyspecific antisera. Since the analysis is performed under nondenaturing conditions it may be possible to identify an antigen by its biological function (e.g., receptor or enzymatic function); the immuno-chemical recognition of the protein combined with characterization by physicochemical techniques then permit a molecular characterization of the protein (Bjerrum 1977).

II. Solubilization

At present the solubilization of membrane proteins for immunochemical investigation is best achieved with nonionic detergents, since the membrane proteins are then maintained under conditions which mimic those in the membrane (Helenius and Simons 1975; Tanford and Reynolds 1976; Bjerrum 1977). With nonionic detergent, membrane solubilization appears to result principally from the replacement of lipid molecules by detergent molecules. The membrane lipids, as well as the amphiphilic "integral" or "intrinsic" membrane proteins, become incorporated into the dominant water-soluble detergent micelles, resulting in membrane disintegration.

Protein-protein interactions are usually not affected during this process and proteins seldom become denatured. Detergent-binding is confined to the apolar surfaces of membrane proteins, and hydrophilic "peripheral" membrane proteins do not significantly bind such mild detergents. Thus a membrane sample solubilized with nonionic detergent comprises a mixture of delipidated, "integral" membrane proteins to which detergent is bound, "peripheral" membrane proteins without bound detergent, and mixed lipid-detergent micelles (Helenius and Simons 1975; Tanford and Reynolds 1976). From these considerations it follows that the presence of nonionic detergents in a protein solution need not affect the hydrophilic molecular regions of proteins. The antigen sites, being primarily confined to such hydrophilic surfaces, remain available for interaction with the antibody (Bjerrum and Bhakdi 1981). This is the principle of which immunochemical analyses of proteins in detergent solution is based.

Although procedures for the solubilization of membranes with nonionic detergents may require modification from system to system, adherence to the following guiding principles generally leads to satisfactory protein solubilization (Helenius and Simons 1975; Tanford and Reynolds 1976; Bjerrum 1977). The protein concentration should not exceed 3–4 mg/ml, and a detergent/protein weight ratio of 3:1 or more is desirable. Solubilization of peripheral membrane proteins may be enhanced by low ionic strength (I < 0.05), pH-changes (pH 3–7 and 8–12) and by the removal of divalent metal ions by addition of the chelators (e.g., EDTA). To minimize formation of disulfide bridges and undesirable enzymatic effects, inhibitors (e.g., iodoacetamide and protease inhibitors such as aprotinin, phenyl methyl-1-sulfonyl fluoride and pepstatin) may be added and the solubilization performed at $0°–4°C$. Sonication of aggregated membrane preparations for short periods (2–10 s at 20,000 Hz) may help to promote solubilization after addition of detergent. As a general rule, solubilized membrane material is recovered as supernatant after ultracentrifugation at 100,000 g for 60 min, although in many instances centrifugation at 40,000 g for 20–30 min will be sufficient to remove aggregates which otherwise appear as smears at the application site in the immunoelectrophoretic analysis.

The Techniques The quantitative immunoelectrophoresis methods are based upon electrophoresis in antibody-containing gel, as devised by Laurell (1965) and Çlarke and Freeman (1968), and in this respect differ from immunoelectrophoretic analysis ad modum Grabar (Grabar and Burtin 1964). Four main features of quantitative immunelectrophoreses are:

1. The resolution is based upon specific recognition of individual proteins by their corresponding antibodies, thereby allowing detailed analysis of individual proteins in crude protein mixtures.

2. The biological activity of the antigen is often retained after immunpre-
   cipitation. This allows further characterization of the proteins, for
   example with respect to receptor functions.

3. The area of the immunoprecipitate is proportional to the amount of
   antigen and inversely proportional to the antibody concentration,
   thereby permitting quantification of proteins.

4. The medium is agarose gel, whose large pores facilitate analysis of hete-
   regeneous large protein complexes (e.g., membrane proteins) and which
   furthermore is easy to handle, thereby facilitating the design of com-
   posite experiments.

Four manuals have been published recently giving all the technical
details of the present state of the technique (Laurell 1972; Axelsen et al.
1973; Axelsen 1975; Svendsen 1979) and the reader is referred to these.
The literature was reviewed extensively up to 1975 by Verbruggen (1975).
Owen and Smyth have described applications in microbiology (1976) and
we have previously discussed the special conditions for the study of membrane proteins by these methods (Bjerrum and Bøg-Hansen 1976b; Bjerrum
1977). The four most useful methods are crossed immunoelectrophoresis,
crossed immunoelectrophoresis with intermediate gel, rocket- and fused
rocket immunoelectrophoresis, all of which will be dealt with in the exercises.

The sensitivity of quantitative immunoelectrophoretic techniques
makes it possible for as little as 10—100 ng of protein to be detected with
the conventional staining technique (Coomassie Brilliant Blue); the sensitivity depending on the antigen-antibody system examined (Laurell 1972;
Axelsen et al. 1973). However, an increase in sensitivity of 10—60-fold can
be obtained by employing enzymatic or autoradiographic methods (see for
example Bjerrum 1977). Thus the membrane antigen may be labeled with
$[^{125}I]$, $[^{14}C]$, $[^{35}S]$ or even $[^{3}H]$ (Norén and Sjöström 1979), or the immuno
plate may be incubated after electrophoresis with specific $[^{125}I]$-labeled
antibodies (Kindmark and Thorell 1972; Plesner 1978). A more general
approach is incubation with enzyme-labeled or $[^{125}I]$-labeled swine or sheep
antibodies against rabbit γ-globulin (the most commonly employed species).
Electrophoretic application at pH 5 of labeled swine antibodies against
rabbit γ-globulins can also be employed (Ramlau and Bjerrum 1977).

Most of these enhancement techniques should in theory allow detection of even smaller amounts of proteins, but are in practice restricted by
the precipitation limit of the antigen-antibody system. Thus, well-defined
precipitates are not formed using antigen amounts less than approx. 0.1—1 ng
(Ramlau and Bjerrum 1977). In cases where a protein is present in these or
in smaller amounts a prefractionation must be performed before the analysis
can take place in order to remove ballast proteins and to obtain a higher

concentration, since a maximum of 50–200 μl of material can be applied in crossed immunoelectrophoresis. In rocket immunoelectrophoresis under certain conditions greater volumes can be applied (Krøll, 1976). A rough impression of the protein material needed for immunoelectrophoretic analysis can be obatined from the precipitation pattern for human erythrocyte membrane proteins shown for example in Picture C, p. 38. In this case about 20 μg of membrane protein was applied, corresponding to approximately 4×10^7 cells (assuming 5.7×10^{-10} mg or protein per hemoglobin-free ghost). The major precipitates spectrin, major "intrinsic" protein (band III) and MN glycoprotein (glycophorin) marked on the figure correspond to proteins present to the extent of about 5×10^5 molecules per cell. Acetylcholinesterase with approx. 6×10^3 molecules per cell could be detected by its enzymatic activity but not by protein staining. The above-mentioned figures suggest that in order to obtain membrane protein precipitates, between 10^7 and 10^8 cells with more than $10^4 – 10^5$ molecules per cell of each protein are necessary (Bjerrum 1977). The situation may change, however, if the protein occurs as a complex with other antigenic proteins which are more frequently represented in the membrane.

Detergent-Immunoelectrophoresis. Removal of the detergent from the amphiphilic membrane proteins results in aggregation, for which reason it is essential to incorporate detergent in the agarose gel during the immunoelectrophoretic analysis. Otherwise the procedures follow those for conventional immunoelectrophoresis (Axelsen et al. 1973) as outlined by Bjerrum and Bøg-Hansen (1976a). However, a few points need attention:

1. Since the agarose gel becomes opaque after addition of the detergent, agarose should first be dissolved and the detergent subsequently added. Otherwise nondissolved agarose may be overlooked and lead to the formation of artifacts.

2. The concentration of neutral detergent in the gel should be above the critical micellar concentration (CMC): 0.1%–1% (w/v) is general suitable. With higher concentrations the viscosity increases, and above 5% (w/v) the gel does not congeal properly. If ionic detergents are to be incorporated, a tritration is needed to determine the optimal concentration; too low concentrations lead to unspecific aggregation and too high concentrations influence the migration of the antibodies.

3. Since the antigen and antibody preparations may contain proteolytic enzymes, it is advantageous to add a protease inhibitor [e.g., aprotinin (Trasylol)] to the gels, 100 KIE/ml gel (Bjerrum and Bøg-Hansen 1976a).

4. Nonionic detergents need not be added to the electrophoresis buffers. If working with ionic detergents, these must be added to the wicks and the buffers.

5. The presence of SDS in a sample may give rise to artifactual nonimmune precipitates when unfractionated anti-serum is used (Gardner and Rosenberg 1969; Bjerrum et al. 1975a). In such cases control experiments are necessary.

The use of different nonionic detergents in the gel may lead to differences in the precipitation pattern. Selective solubilization or varying dissociative effects of the detergents on protein complexes may be responsible for changes in the number of precipitates, whilst variations in the electrophoretic migration may be due to differences in size of the different detergent micelles, which bind to the proteins and reduce the average charge density of the proteins. The appearance of asymmetric precipitates or even double peaks indicates molecular heterogeneity of the solubilized antigen. As monomerization of membrane proteins by solubilization is seldom complete, membrane precipitates often appear with some tailing in the precipitation pattern. Differences in the relative area below the precipitates reflect the amount of protein applied, but also differences in the solubilization efficiency of the detergents in question. Furthermore, variations in size and migration velocity of the protein-detergent complexes may change the area delimited by the precipitate.

At detergent concentrations below the CMC aggregation of "integral" membrane proteins often occurs. This gives rise to large irregular, blurred precipitates which hamper the immunophoretic analysis (Bjerrum and Bøg-Hansen 1976a). Furthermore, as a result of the dissociation of the detergent "micelle" from the "intrisinc" proteins, the latter have a higher migration velocity, due to their increased average charge density. Normally the proteins will be delipidated by the action of the detergent, for which reason proteins whose antigenicity is dependent on the presence of lipid may lose their reactivity.

Quantitative Immunoelectrophoresis of proteins in the presence of ionic detergents is possible (Bjerrum and Bhakdi 1981), but is hampered by several factors: fewer precipitates are seen with SDS due to denaturation and, furthermore, the addition of the charge of SDS to the protein molecules minimizes the resolution in the first-dimension electrophoresis. In a similar manner SDS binds to antibody molecules during second-dimension immunoelectrophoresis, causing electrophoretic removal of the immunoglobulins from the agarose gel. On the other hand, lowering of the detergent concentration gives rise to aggregation phenomena as already described for nonionic detergents. The effects of SDS can be partially counteracted by the introduction of nonionic detergents into the gels (Bjerrum et al. 1975a). Thus a membrane sample can be solubilized with a ionic detergent and then analyzed in gels containing nonionic detergent (1% or more). The protein-bound ionic detergent is then replaced with nonionic detergent during the immunoelectrophoresis with concomitant antigenic renaturing. In this way it is possible to perform polyacrylamide gel electrophoresis with SDS for the first dimension and combine it directly with a second-dimension immunoelectrophoresis (Converse and Papermaster 1975; Chua and Blomberg 1979).

Neutral zwitterionic detergents carry both positively and negatively charged groups but their net charge in the pH region of 3–11 is essentially zero. Examples are Empigen (Allen and Humphries 1975) and the sulfobetaine series (Goenne and Ernst 1978). The advantage provided by these detergents is their high solubilizing efficiency without affecting the electrophoretic migration of antigens and antibodies as a result of binding. However, these detergents are still rather denaturing in their action.

The General Strategy for membrane protein characterization and isolation by means of quantitative immunoelectrophoretic methods normally involves the following steps: (1) Production of a polyspecific antibody; (2) establishment of a reference pattern by crossed immunoelectrophoresis of solubilized membrane protein; (3) identification of the biologically active protein in the precipitation pattern (Bjerrum et al. 1981); (4) molecular characterization of the protein by various biochemical and physicochemical techniques combined with the immunoelectrophoretic analysis; (5) testing of the possible fractionation methods on an analytical scale by means of immunoelectrophoretic "table-top" techniques (Bjerrum 1978); (6) final isolation on a large scale with subsequent molecular characterization regarding aspects not covered by the foregoing techniques.

III. Equipment, Reagents and Solutions

A. Equipment

Watercooled Electrophoresis Apparatus. Basically it consists of a closed chamber containing two buffer vessels equipped with electrodes and a cooled surface equipped with supports for wicks or agarose gel bridges. A suitable electrophoresis apparatus in "Danish design" is shown in Fig. 1. This type has a cooled surface of 22 × 12 cm which provides room for three 7 × 10 cm plates. For the described exercises three of such apparatuses are necessary. Cooling water at a flow rate of 1 l/min may come from the water tap or be circulated by a cooling thermostat adjusted to 15°C.

A

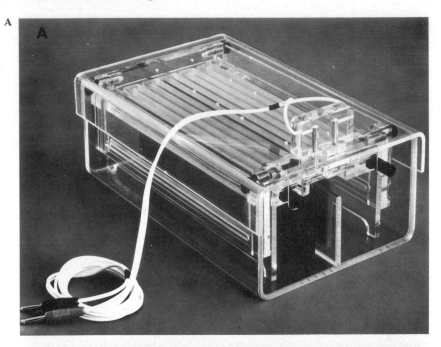

B

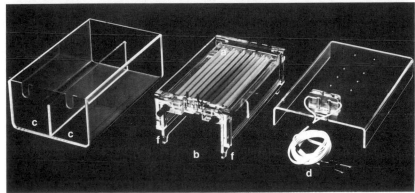

Fig. 1. Electrophoresis apparatus for electroimmunoprecipitation. In **A** it is assembled and in **B** the main parts are shown. *a* Cooled surface; *b* electrodes; *c* electrode vessels; *d* lid with electric connectors; *f* support with moulds for agarose gel connections. Producer: Holm Nielsen Aps., DK–2460, Søborg, Denmark

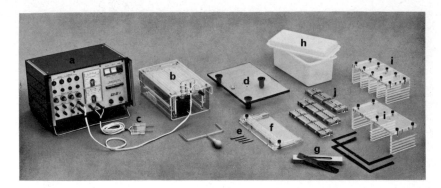

Fig. 2. Apparatus and accessories for electroimmunoprecipitation. *a* Power supply; *b* electrophoresis apparatus; *c* test probe for measuring the potential gradient; *d* adjustable glass table for casting gels; *e* gel punchers; *f* adjustable template for punching sample wells; *g* long razor blades for cutting and transferring gel slabs; *h* vessel for washing and staining; *i* racks for holding glass plates; *j* slit formers for production of narrow application wells

Power Supply. It should have a capacity of 300 V and 100–200 mA D.C. and contain at least three channels with individual regulation (see a in Fig. 2).

Test Probe. (See c in Fig. 2). In connection with a voltmeter this is used to measure the potential gradient directly in the gel when the current has been switched on. This can be done through the lid of the electrophoresis apparatus if it is furnished with a suitable pair of holes.

Thermostated Water Bath. This should maintain a temperature of 56°C and have room for three 250 ml Erlenmayer flasks containing stock solutions of melted agarose, a rack for test tubes containing agarose + antibody and a 250 ml measuring cylinder where the pipettes for the agarose distribution can be kept warm.

Test Tubes. These should be equipped with lips: 25 ml and 5 ml.

Pipettes. For pipetting agarose: 25, 10, and 5 ml graduated glass pipettes. For pipetting antibodies: 200, 250, 200, and 500 μl pipettes (any type). For application of samples: 25 μl or 50 μl Hamilton syringe or double constriction micropipettes (5 μl and 10 μl).

Glass Plates. The following sizes are used: 100 × 100 mm, 100 × 70 mm, 100 × 50 mm and 70 × 50 mm. The thickness should be between 1 and 2 mm.

Diamond Pencil for marking the glass plates.

Leveled Glass Table. To ensure equal thickness of the gel on casting, a table (d in Fig. 2) or just a large glass plate should be leveled in combination with a precision spirit level.

Gel Punchers with diameters of 2.5 mm and 4.0 mm (e in Fig. 2).

Template for making a row of holes for fused rocket immunoelectrophoresis (f in Fig. 2).

Long Razor Blades (2 X 15 cm) for cutting and transferring agarose gel slabs fron one plate to another (g in Fig. 2).

Brass Bar (7 X 1 X 1 cm) for casting intermediate gels.

Wicks. Whatman No. 1 filter paper cut in appropriate sizes.

Vessels for washing and staining solutions (h in Fig. 2).

Rack for holding glass plates during washing and staining (i in Fig. 2).

Plate of Wood or Glass (60 X 30 X 1 cm) for pressing the gels. Pressure: $1-2$ g/cm^2.

Filterpaper for absorbing the water during the pressing procedure. It may be cut in the same size as the immuno plates.

Fan or Hairdryer for drying the gels before staining.

B. Reagents

Triton-X-100 (scintillation grade, BDH), Sodium deoxycholate (für Microbiologie, Merck) N-Cetyl-N,N,N,-trimethylammonium bromide (Merck), wheat germ lectin and wheat germ lectin-Sepharose (*Macro Beads*) (Pharmacia Fine Chemicals, Uppsala). Wheat germ lectin coupled to *normal* sepharose 4B (5 mg/ml). The latter gel was prepared from CNBr-activated Sepharose (Pharmacia Fine Chemicals, Uppsala) using the directions given by the manufacturer.

Markers: Bromophenol-blue marker [human serum plus 2 mg/ml of bromophenol-blue (Merck)] and hemoglobin marker (human hemolysate: dilution of washed erythrocytes 1:9 (v/v) with H_2O, centrifugation at 25,000 g for 20 min).

Antigens. Human erythrocyte membranes isolated from outdated bank blood according to Dodge et al. (1963) (they contain residual hemoglobin). Human albumin (Reinst, Behringwerke).

Antibodies. Rabbit immunoglobulins against human erythrocyte membrane proteins (code A 104), human albumin (code 10-001) and human hemoglobin (code A 118). (All from DAKO immunoglobulins, Copenhagen F, Denmark). Rabbit immunoglobulins to bovine erythrocyte membrane proteins were produced according to Bjerrum (1975).

C. Solutions

Buffer for Electrode Vessels and Gels. A stock solution of 1 M glycine and 0.38 M Tris (7–9 Sigma), pH 8.7 should be diluted 1 + 9 before use.

Gels. 1% (w/v) Litex agarose Type HSA, M_r = 0.13 (Litex, Glostrup, Denmark) in above mentioned buffer containing (a) 0.5% (v/v) Triton X-100; (b) 0.5% (v/v) Triton X-100 + 0.2% (w/v) deoxycholate; (c) 0.5% (v/v) Triton X-100 + 0.125% (w/v) cetyltrimethylammonium bromide.

Staining Solution. Coomassie Brilliant Blue 0.5% (w/v) in ethanol (96%), glacial acetic acid, distilled water (45%, 10%, 45%). The stain is dissolved in the acidic aqueous ethanol with stirring overnight. After filtration it is ready for use.

Destaining Solution. Ethanol (96%), glacial acetic acid, distilled water (45%, 10%, 45%).

IV. Experimental Procedures

A. Solubilization

Washed human erythrocyte membranes (ghosts) which have been stored frozen, are solubilized by mixing the membranes with 0.038 M Tris and 0.1 M glycine buffer (pH 8.7) and 20% (v/v) Triton X-100 so that the final protein concentration is 2 mg/ml and the Triton X-100 concentration is 1.0%, total volume 2 ml. The mixture is then sonicated at 20.000 Hz for 2 × 5 sec on ice and centrifuged at 20,000 r.p.m. (40,000 g) for 1 h at 5°C (rotor SS-12, Sorval RC-5B centrifuge (Bjerrum and Bøg-Hansen 1976b). The supernatant is kept on ice and used for the immunoelectrophoretic experiments.

B. Histochemical Staining

Principle. Many biologically important proteins such as enzymes can be identified in immunoprecipitation experiments by means of their biological activity. Precipitation of enzymes by antibodies does not normally abolish enzyme activity and histochemical color reactions can therefore be used for specific staining of immunoprecipitates in gels.

Purpose. Demonstration of esterase activity associated with an immunoprecipitate.

Performance:

1. Crossed immunoelectrophoresis plates are pressed and then dried in a stream of cold air.
2. The plates are immersed in an incubation medium of 0.2 M Tris-HCl buffer (pH 7.5), 0.2 mg/ml enzyme substrate (α-naphtyl acetate, Sigma) previously dissolved in acetone as a 1% (w/v) solution and a pinch (about 0.5 mg per ml) of diazonium coupling salt (fast red TR, Sigma). Hydrolysis of the naphtyl ester results in an aromatic alcohol which couples to the diazonium salt and forms an insoluble azo dye (Brogren and Bøg-Hansen 1975; Bjerrum et al. 1975b).

C. Crossed Immunoelectrophoresis (Laurell 1965)

Principle. In crossed immunoelectroporesis (a.m. Clarke and Freeman 1968) the proteins are separated electrophoretically in an agarose gel. All the separated proteins are then run at right angles to the first direction into a gel containing antibodies against the proteins. Every protein for which a corresponding antibody is present gives rise to a precipitation peak. The area enclosed by the precipitate is proportional to the amount of antigen. The resolving power of the crossed immunoelectrophoresis is high, making it possible to quantify 20–30 proteins in one run. The precision is 2%–18% depending on the antigen (Axelsen et al. 1973).

Purpose. To investigate a complex protein mixture of human erythrocyte membrane proteins solubilized in Triton X-100.

Plan 1. Crossed immunoelectrophoresis

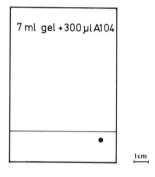

This and the following plans are shown in reduced scale. (The bar indicates 1.0 cm.) Before an experiment is carried out the plan should be redrawn in full scale on graph paper and serve as a template for the positions of the application wells, intermediate gels etc., as the cuts of the first dimension gels.

Performance (see Plan 1)

First dimension

1. Clean a 10×10 cm glass plate with a paper tissue moistened with ethanol.

2. Place the plate on a level surface.

3. Mount a 25 ml test tube in the waterbath and pipet 15 ml of agarose into the tube. Pour the agarose onto the glass plate.

4. Punch wells according to Plan 1.

5. Apply antigen; note that bromophenol-blue stained serum is applied in the well at the top!

6. Clean the electrophoresis apparatus. Fill the electrode vessels with buffer and ensure that the electrodes are covered and the level in the two vessels is the same.

7. Place the glass plate (the gel uppermost) on the cooling surface of the electrophoresis apparatus. Connecting bridges, for example 8 layers of Whatman no. 1 filter paper moistened with the buffer, are then established. The wicks should overlap the edges of the slide by approximately 1 cm.

8. Switch on the current and adjust the voltage across the gel to about 10 Volt per cm using the test probe and the voltmeter. At pH 8.7 the proteins will migrate to the positive electrode. Continue the electrophoresis until the front of the light blue marker (the albumin) has migrated 4.5 cm (ca. 50 min). In the presence of nonionic detergent the free dye is retarded and appears cathodically to the albumin spot.

Second dimension

9. Clean a 7 X 10 cm glass plate. Moisten a paper tissue with molten agarose and coat the surface of the plate with a thin agarose film. Allow to dry. The coating helps the antibody containing gel to adhere to the slide, but is not necessary to perform.

10. Cut out and transfer first dimension gel strip to the coated plate (see Plan 1).

11. Mount a 25 ml test tube in the waterbath and pipette 7 ml of agarose into the tube. When the agarose is cooled down to 56°C, add the antiserum. Mix thoroughly taking care to avoid bubbles or foam and pour the antibody-containing gel onto the slide. Allow to set.

12. Place the plate on the apparatus, and connect to buffer reservoir by 4 layers of filter paper.

13. Adjust the potential gradient to 2.0 V/cm and allow to run overnight.

14. With a polyspecific immunoglobulin fraction of antiserum, washing of the plate is necessary to obtain a clean background:
   a) Press under 5 layers of filter paper for 5 min, renew the 4 upper layers of paper and press for another 5 min.
   b) Rinse in 0.1 M NaCl, 15 min.
   c) Press under filter paper, 10 min.
   d) Rinse in distilled water, 15 min.
   e) Press under filter paper, 10 min.
   f) Because of the histochemical staining the plates should be dried in a stream of cold air. (Normally plates can be dried in warm air.)

g) Perform immunohistochemical staining for esterase activity.

h) Rinse and dry again.

i) Stain for 5–10 min in Coomassie Brilliant Blue, destain and dry.

Evaluation

1. Place the plate on white paper, gel side down.

2. How many precipitates can be seen?

3. Which precipitates can be recongized when compared with the reference of Picture C on p. 38 (write on the glass side of the plate). Is this identification safe?

4. Point out proteins with electrophoretic heterogeneity.

5. Any technical errors?

6. What happens if the nonionic detergent is omitted from the gel?

D. Rocket Immunoelectrophoresis (Laurell 1966)

Principle. When electrophoresis of an antigen is performed in an agarose gel containing the corresponding antibody, a long rocket-like immunoprecipitate develops. The length of the rocket is porportional to the antigen amount.

Rocket immunoelectrophoresis gives rapid and precise identification and quantification of even minor amounts of proteins. The standard deviation on double determinations is 1%–3%, and as little as 4 nanogram of albumin can be detected. The electrophoretic run can be completed in 2–16 h, depending on the electrophoretic migration velocity of the antigen and the voltage gradient employed.

Purpose. Measurement of the albumin content of washed erythrocyte membranes.

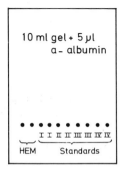

Apply 10 µl   **Plan 2.** Rocket immunoelectrophoresis

Performance (see Plan 2)

1. Prepare 4 albumin standards diluted 1 + 5, 1 + 10, 1 + 20, 1 + 50 with 0.154 M NaCl from an albumin stock solution of 0.05 mg/l.
2. Rinse a 7 × 10 cm glass plate with alcohol.
3. Mix 10 ml agarose with 5 μl anti-albumin. Pour the gel onto the plate.
4. Punch wells according to Plan 2.
5. Apply 10 μl of standards and samples as indicated on Plan 2. Use double application of the four standards, and apply the standard with the lowest concentration first. Use a double constriction pipette, the same for all applications.
6. Adjust the potential gradient in the gel to 2.0 Volts per cm and allow to run overnight.
7. For washing and staining, see under Crossed Immunoelectrophoresis. (No esterase staining.)

Evaluation. Calculate the albumin content in the solubilized membrane sample. The height of the rockets are measured with an accuracy of 0.5 mm by placing the plates on top of graph paper. A linear or near-linear relationship is obtained by plotting the mean heights of standard rockets against the amount of antigen applied to the wells. By interpolation on the curve unknown samples can be quantified. Rockets used for quantification should measure between 5 and 50 mm.

E. Analytical Affinity Electrophoresis with Lectin
(Bøg-Hansen 1973; Bøg-Hansen, Bjerrum and Ramlau 1975)

Principle. Lectins are proteins with affinity for various carbohydrates. Interactions between lectins and glycoproteins can be studied by means of immunoprecipitation techniques such as crossed affino-immunoelectrophoresis. The latter is a modification of crossed immunoelectrophoresis with intermediate gel in which a lectin coupled to agarose is mixed into the intermediate gel instead of antibodies. Antigens, which interact with the lectin, revealed by the disappearance of the corresponding immunoprecipitate from the antibody-containing gel, are regarded as glycoproteins. In control experiments 5% (w/v) of the relevant carbohydrate can be incorporated into the lectin-containing gel. In this case the specific interaction between lectin and carbohydrate-containing antigens can be inhibited (Bøg-Hansen 1973).

Alternatively, lectin can be incorporated into the first-dimension gel and interacts with the antigens during the first-dimension electrophoresis giving

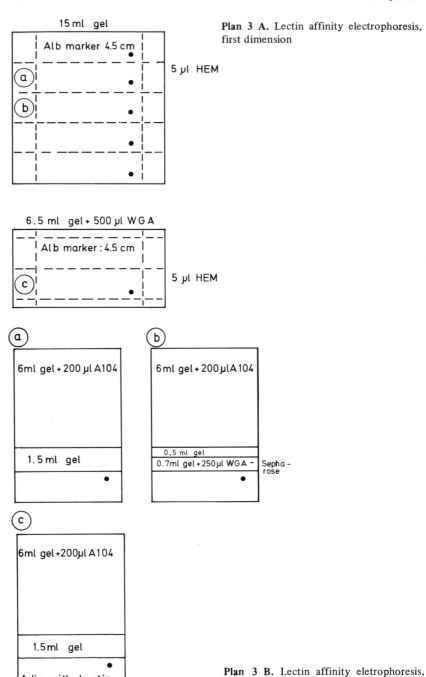

Plan 3 A. Lectin affinity electrophoresis, first dimension

Plan 3 B. Lectin affinity eletrophoresis, second dimension

rise to (a) precipitation in the first dimension gel, (b) changed position and (c) changed shape of the precipitate (Bøg-Hansen et al. 1975).

Purpose. To show different forms of affinity electrophoresis with wheat germ lectin (agglutinin) WGA, for identification of N-acetyl glucosamine-containing human erythrocyte membrane proteins and for prediction of preparative affinity chromatography. (Compare with experiment I.: Fused Rocket Immunoelectrophoresis.)

Performance (see Plan 3)

1. The participant performs three experiments as outlined on Plan 3.
2. Two plates are used for the first dimension: (1) 10 X 10 cm and (2) 10 X 5 cm. Agarose and agarose with lectin are cast on these plates. WGA is provided as a 5 mg/ml solution in water. Use 500 µl.
3. The sample for investigation consists of Triton-solubilized human ery-throcyte membrane proteins. Apply samples 3 X 5 µl according to Plan 3, with 2 samples in ordinary agarose gel (a + b) and one in WGA-containing gel (c). The other wells may also be filled in the same way as a reserve.
4. First-dimension electrophoresis is run until bromophenol blue-marked albumin has migrated 4.5 cm on both plates (from front of well to front of light blue spot). Applied voltage, 10 V/cm.
5. The 3 first dimension gels are transferred to 7 X 10 cm plates and with use of the brass bar 3 intermediate gels are moulded according to Plan 3:
    a) 1.5 ml gel,
    b) 0.7 ml gel + 250 µl WGA-Sepharose 4B (33% gel slurry; mix before use), followed by 0.5 ml blank gel.
    c) 1.5 ml gel.
6. The second dimension gel contains 200 µl antiserum against human ery-throcyte membrane proteins (A104). The second electrophoresis is run at 2 V/cm, and the plates are processed normally.

Evaluation

1. Is any binding of proteins to lectin apparent?
2. Do the two different techniques give identical results?
3. Compare the results obtained with those from experiment I: the affinity chromatography experiment on WGA.
4. Why is the blank intermediate gel inserted?

F. Crossed Immunoelectrophoresis with Intermediate Gel.
Comparison of Antisera (Svendsen and Axelsen 1972)

Principle. This is a modified crossed immunoelectrophoresis. After the first-dimensional separation of antigens, an antibody-containing gel is placed between the first dimension gel and the second dimension gel containing the reference antiserum. During the second dimension electrophoresis the antigens will migrate through the intermediate gel and thereafter through the reference gel. Depending upon the amount of antigen and the antibody titers of the intermediate gel, the antigens will be (a) completely retarded, (b) partially retarded or (c) unretarded by immunoprecipitation in the intermediate gel. This is reflected in the reference precipitate pattern by (a) the absence of a precipitate, (b) fusion of reference and intermediate gel precipitates or (c) no changes in the reference precipitate pattern as determined by comparing the test plate with an obligatory control plate without antibody in the intermediate gel (Axelsen et al. 1973).

Purpose. To identify the precipitate corresponding to hemoglobin in the reference pattern of Triton-solubilized human erythrocyte membrane proteins and to compare an antiserum against human erythrocyte membrane proteins with an antiserum against bovine erythrocyte membrane proteins.

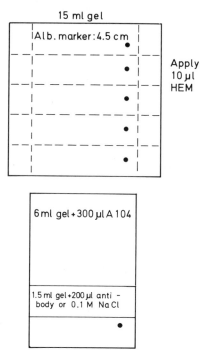

Plan 4. Crossed immunoelectrophoresis with intermediate gel

Performance (see Plan 4)

1. Three immunplates (A, B, C) are prepared. $3 \times 10 \, \mu l$ of Triton-solubilized human erythrocyte membrane protein is used as antigen.

2. The intermediate gel in plate A (control plate) consists of $200 \, \mu l$ 0.1 M NaCl mixed with 1.5 ml gel. In plate B: $200 \, \mu l$ anti-hemoglobin antibody + 1.5 ml gel, and in plate C: $200 \, \mu l$ anti-bovine erythrocyte membrane protein antibody + 1.5 ml gel.

3. Reference antibodies on all three plates (A + B + C): $300 \, \mu l$ antierythrocyte membrane protein antibodies + 6 ml gel.

Evaluation

1. How many different antibody-specificities could be found in the heterologous antiserum?

2. Could the corresponding proteins be identified?

3. How can antibody titers be expressed in this technique?

4. Why are some of the reference precipitates of the control plate present in the intermediate gel?

5. Does the experiment provide proof of immunlogical similarity between human proteins and animal proteins?

6. Were there any technical errors?

G. Determination of Topography of Membrane Antigens

Principle. Intact and lyzed membrane structures are added to an antiserum containing antibodies against one or several membrane proteins. Membrane proteins on the surface of the structures will be easily accessible to the corresponding antibodies, whereas proteins located on the inner side of the membrane can only react with their corresponding antibodies when the structures are lyzed (Howe and Lee 1969). The decrease in antibody titers can thereafter be determined.

Performance (see Plan 5)

1. The absorbed antisera have been prepared as follows (Bjerrum et al. 1975b): 100 ml blood (in heparin) is washed $\times$ 5 with 0.154 M NaCl with centrifugations of $1000 \, g \times 5$ min. A final centrifugation of $500 \, g \times 15$ min will result in pellet containing intact packed erythrocytes comprising about 66% of the volume (the remainder is saline).

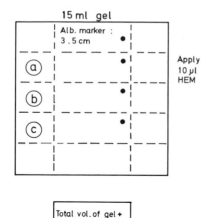

15 ml gel

Alb. marker :
3.5 cm

(a)

(b)

(c)

Apply
10 µl
HEM

Total vol. of gel +
antibodies =
3.7 ml

Plan 5. Crossed immunoelectrophoresis.
Membrane topography

12 ml packed erythrocytes are mixed with 1.0 ml anti-erythrocyte membrane antibody (dialyzed against 0.154 M NaCl) and 5.0 ml 0.154 M NaCl. The mixture is incubated at room temperature for 30 min with occasional shaking. After incubation the cells are removed by centrifugation for 1000 g X 5 min and the supernatant containing the antiserum (diluted 1:10) is withdrawn.

12 ml packed erythrocytes are lyzed by repeated freezing and thawing (3 X). 1.0 ml dialyzed anti-erythrocyte membrane antibody is added and the mixture is incubated as above (antiserum dilution ≈ 1:10). This antiserum is employed to check that the insolubilized membrane antigens are able to bind all antibodies.

Dialyzed unabsorbed antiserum serves as a control.

2. Human erythrocyte membrane proteins solubilized in 1% (v/v) Triton X-100 are employed as antigens. Crossed immunoelectrophoresis in gels containing 0.5% Triton X-100 is performed as previously described, using Plan 5. Three 5 X 7 cm immunoplates (a, b, c) are run using 10 µl solubilized erythrocyte membranes as antigen and:

a) 150 µl anti-erythrocyte membrane antibody (unabsorbed)
b) 1500 µl anti-erythrocyte membrane antibody absorbed with intact cells
c) 1500 µl anti-erythrocyte membrane antibody absorbed with lysate.

3. The plates should be stained for esterase activity. (Remember to dry the plate with cold air.)

Evaluation

1. How many of these proteins were localized on the external surface?

2. If any of the membrane proteins span the membrane with antigenic determinants on both sides, how would this be detected in the present system?

3. Propose a method to check that an identical dilution of the antibody has been used in the test plate and the control plate.

H. Charge-Shift Crossed Immunoelectrophoresis
(Bhakdi, Bhakdi-Lehnen and Bjerrum 1977)

Principle. The original technique of charge-shift electrophoresis was developed by Helenius and Simons (1977). The method is based on the observation that significant alterations in the electrophoretic migration of amphiphilic proteins occur when they bind detergent micelles of different charge. The experimental system therefore involves three parallel electrophoresis of identical protein samples; in the first system, which serves as the control, electrophoresis is performed in agarose gels which contain the nonionic detergent, Triton X-100. In the second system the anionic detergent sodium deoxycholate is added which contributes a negative charge at pH 8.7 to the Triton micelles, into which it becomes incorporated. In the third sytem, the Triton micelles are rendered positive in charge through the addition of low concentrations of the cationic detergent cetyltrimethylammonium bromide. Any resulting significant bidirectional "charge-shifts" exhibited by a protein therefore arise through its Triton-binding properties, and not through the selective interaction with deoxycholate or cetyltrimethylammonium bromide.

If a protein is hydrophilic and does not bind Triton X-100, its electrophoretic migration in the three electrophoretic systems will remain completely constant. If a protein binds Triton micelles, its electrophoretic migration will differ in all three systems because of the differences in net charge of the migrating protein-detergent complexes. Monodirectional shifts due to selective binding of one of the charged detergents may occur.

Purpose. Detection, in the crossed immunoelectrophoresis, of the amphiphilic proteins present in a Triton X-100 extract of human erythrocyte membranes.

3 different 1.dimension plates: Plate 1: Tx -100 ;
Plate 2: Tx-100+DOC; Plate 3: Tx-100+CTAB

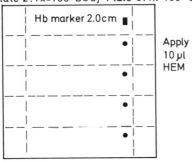

2.dimension gel contains Tx - 100 , only.

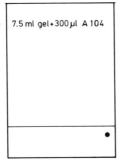

Performance (see Plan 6)

1. The following buffer systems are used: (A) 0.5% (v/v) Triton X-100 added to agarose, no detergent is necessary in the buffers, (B) 0.5% (v/v) Triton X-100 plus 0.2% (w/v) deoxycholate present in the agarose *and* buffers, (C) 0.5% Triton X-100 plus 0.0125% (w/v) cetyltrimethyl-ammonium bromide in gels *and* buffers. 10 × 7 cm plates are used. See Plan 6.

2. 10 μl Triton-solubilized human erythrocyte membrane proteins are applied in wells in the respective gels and are then simultaneously electrophoresed at 10 V/cm in the respective buffer systems until a hemoglobin marker (10 μl), which is hydrophilic has migrated 20 mm. The distance of the marker from the front of the starting well is measured to the front of the marker spot in the gels. The marker molecule is allowed to migrate an equal distance in all three gels. Use of a slit makes it easier to measure the distance. [For later evaluation the marker gels may be fixed in a picric acid solution (see Laurell 1972).]

3. After completion of first-dimension electrophoresis the agarose strips are sectioned and transferred to 7 × 10 cm glass plates, and 7.5 ml agarose containing 0.5% Triton X-100 and 300 μl antibodies against human erythrocyte membrane is cast as in normal crossed immunoelectrophoresis. The buffer for second-dimension electrophoresis does not contain detergent. The buffers of the electrophoresis apparatus containing the charged detergents should be exchanged, so that they are ready for other experiments.

Evaluation. The positions of the immuno-precipitates obtained from the three parallel immunoelectrophoreses are compared with one another. Charge-shifts are defined as any difference in electrophoretic migration of a protein in the presence of the charged detergents, compared to its migrational position in the presence of Triton X-100 alone. The charge-shifts are expressed in mm distance, measured from the positions of the precipitation peaks in the case of bell-shaped symmetrical immunoprecipitates, or from the median of the immunprecipitate if it is broader and lacks a clearly defined peak. The position of the precipitate formed in the Triton system serves as the basis for measurements.

Under the conditions described a bidirectional shift of more than ± 5 mm is indicative of an amphiphilic protein (Bhakdi et al. 1977). Can such proteins be identified here?

I. Fused Rocket Immunoelectrophoresis (Svendsen 1973)

Principle. This technique is a modified version of rocket immunoelectrophoresis. Small volumes of fractions obtained in separation experiments are transferred to a row of sample wells in an antibody-free agarose gel, where the applied samples are allowed to diffuse into the gel for 30–60 min. The proteins are then forced by electrophoresis into an antibody-containing gel. The result is that elution profiles for individual proteins are obtained in one single experiment.

Purpose. To investigate the separation of Triton-solubilized erythrocyte membrane proteins achieved by:
A. affinity chromatography on wheat germ agglutinin;
B. sucrose-density gradient centrifugation;
C. gel filtration experiment.

Affinity Chromatography of 1.5 mg Triton X-100-solubilized human erythrocyte membrane protein was performed on 2.5 ml WGA-Sepharose MB column (Pharmacia Fine Chemicals, Uppsala). The column was a 5 ml plastic syringe. Each fraction (1 ml) was eluted by hand with 0.1% (w/v) Triton X-100 in 0.1 M glycine, 0.038 M Tris (pH 8.7) and then with 10% (w/v) of N-acetylglucosamine in the above-mentioned buffer.

Sucrose-Density Gradient Centrifugation. 1.5 mg of Triton X-100-solubilized human erythrocyte membrane proteins, 0.116 M phosphate buffer (pH 7.4), 1% (v/v) Triton X-100 were applied on a 4 cm long Triton X-100-containing [0.5% (v/v)] linear sucrose gradient [4 ml tube, 10%–35% (w/v)]. Ultracentrifugation at 40,000 rpm (1.4 10^5 × g_{av}) was performed for 18 h. Fractions were obtained by draining the tubes from the bottom.

Gel Filtration Experiment. 10 mg of Triton X-100-solubilized human erythrocyte membrane proteins [0.1 M glycine, 0.038 M Tris (pH 8.7) and 1% (v/v) Triton X-100] were applied on a 300 ml column (length 95 cm) of ACA 34 (MW range 20,000–350,000, LKB, Bromma, Sweden) equilibrated with 0.05% (v/v) Triton X-100 in the above mentioned buffer containing 15 mM NaN_3. Elution was performed with 8 ml/h and fractions were collected every 45 min. Fractions 11–45 are examined.

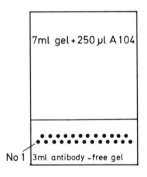

Apply 10 μl aliquots
Diffusiontime : 1 hr   **Plan 7.** Fused rocket immunoelectrophoresis

Performance (see Plan 7)

1. Pour the antibody-free gel of 3 ml against a brass bar. After the gel has congealed, the bar is cut free.

2. Pour the antibody containing gel.

3. Punch wells with 4.0 mm diameter puncher using the template.

4. Apply aliquotes of the fractions successively in the wells (the pipette need not be washed between applications); regarding volume of samples, see Plan 7:

5. Leave to diffuse for 1/2—1 h on the bench under cover.

6. Electrophoresis: 2.0 V/cm overnight.

7. Press etc. as for Crossed Immunoelectrophoresis.

Evaluation

1. Try to correlate the precipitates to the reference pattern obtained in crossed immunoelectrophoresis.

2. Propose methods to ensure a reliable identification.

3. Were any proteins isolated by the fractionation procedures employed?

4. Is it possible to determine the size of molecules by direct observations on the precipitates?

5. What are the advantages of the method?

6. Suggest cases where the method would give misleading results.

J. An Atlas of Typical Results

The label in the upper left corner refers to the various experiments performed. The following precipitates of the reference pattern have been
identified:

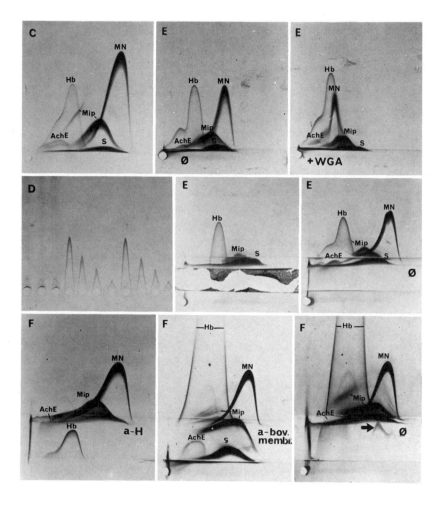

MN-glycoprotein or glycophorin *(MN)*, the spectrins *(S)*, major "intrinsic" protein or band III protein is present in two precipitates: Mip (these correspond to complexed and uncomplexed Mip and the dissociation appears to various degrees), hemoglobin *(Hb)* and acetylcholinesterase *(AchE)*. When possible these proteins are indicated on the plates. See Bjerrum and Bøg-Hansen (1976b) for further details.

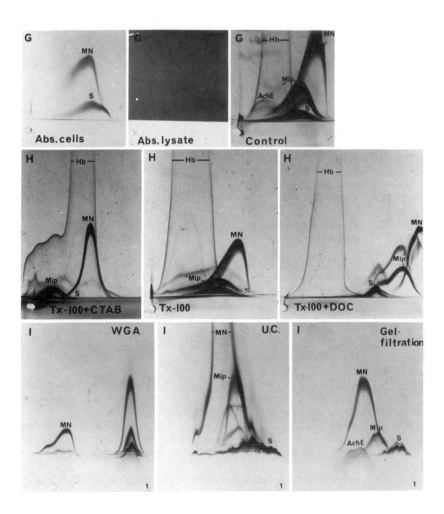

Acknowledgments. At the Protein Laboratory a row of workshops on electroimmuno-chemical precipitation methods has been given during the past decade from which the exercises here presented have their roots. The demonstration plates have been made by Mrs. K. Norup Jensen. Thanks are due to Pharmacia Fine Chemicals for delivering WGA coupled to Sepharose 4B and to Holm Nielsen Aps, DK–2460, Søborg, Denmark, for the Figures 1 and 2.

References

Allen JC, Humphries C (1975) The use of zwitter-ionic surfactants in the agarose chro-matography of biological membranes. FEBS Lett 57:158–162

Axelsen NH (1975) Quantitative immunoelectrophoresis. New developments and appli-cations. Universitetsforlaget, Oslo, alias Scand J Immunol 4: Suppl 2

Axelsen NH, Krøll J, Weeke B (1973) A manual of quantitative immunoelectrophoresis. Methods and applications. Universitetsforlaget, Oslo, alias Scand J Immunol 2: Suppl 1

Bhakdi S, Bhakdi-Lehnen B, Bjerrum OJ (1977) Detection of amphiphilic proteins and peptides in complex mixtures. Charge-shift crossed immunoelectrophoresis and two-dimensional charge-shift electrophoresis. Biochim Biophys Acta 470:35–44

Bjerrum OJ (1975) Quantitative immunoelectrophoresis for comparative analysis of membrane proteins from various mammalian species. Int J Biochem 6:513–519

Bjerrum OJ (1977) Immunochemical investigation of membrane proteins. A methodo-logical survey with emphasis placed on immunoprecipitation in gels. Biochim Biophys Acta 472:135–195

Bjerrum OJ (1978) Crossed hydrophobic interaction immunoelectrophoresis: an analy-tical method for detection of amphiphilic proteins in crude mixtures and for pre-diction of the result of hydrophobic interaction chromatography. Anal Biochem 90:331–348

Bjerrum OJ, Bhakdi S (1981) Detergent-immunoelectrophoresis of membrane proteins. In: Axelsen N (ed) General principles in Manual of immunodiffusion and electro-immuno-precipitation methods. Pergamon Press, London (in press)

Bjerrum OJ, Bøg-Hansen TC (1976a) In: Maddy AH (ed) Biochemical analysis of mem-branes. Chapman and Hall, London, pp 378–426

Bjerrum OJ, Bøg-Hansen TC (1976b) The immunochemical approach to the characteri-zation of membrane proteins. Human erythrocyte membrane proteins analyzed as a model system. Biochim Biophys Acta 455:66–89

Bjerrum OJ, Lundahl P (1974) Crossed immunoelectrophoresis of human erythrocyte membrane proteins. Immunoprecipitation patterns for fresh and stored samples of membranes extensively solubilized with non-ionic detergents. Biochim Biophys Acta 342:69–80

Bjerrum OJ, Bhakdi S, Bøg-Hansen TC, Knüfermann H, Wallach DFH (1975a) Quanti-tative immunoelectrophoresis of proteins in human erythrocyte membranes. Ana-lysis of protein bands obtained by sodium dodecylsulphate-polyacrylamide-gel-electrophoresis. Biochim Biophys Acta 406:489–504

Bjerrum OJ, Lundahl P, Brogren C-H, Hjerten S (1975b) Immunoabsorption of mem-brane-specific antibodies for determination of exposed and hidden proteins in human erythrocyte membranes. Biochim Biophys Acta 394:173–181

Bjerrum OJ, Ramlau J, Bock E, Bøg-Hansen TC (1981) Immunochemical identification and characterization of membrane receptors. In: Jacobs S, Cuatrecases P (eds) Membrane receptors: Methods for purification and characterization. Chapman and Hall, London

Bøg-Hansen TC (1973) Crossed immuno-affino electrophoresis. A method to predict the result of affinity electrophoresis. Anal Biochem 56:480–488

Bøg-Hansen TC, Bjerrum OJ, Ramlau J (1975) Detection of biospecific interaction during first dimension electrophoresis in crossed immunoelectrophoresis. Scand J Immunol 5: Suppl 2, 141–147

Brogren C-H, Bøg-Hansen TC (1975) In: Axelsen NH (ed) Quantitative immunoelectrophoresis. New developments and applications. Universitetsforlaget, Oslo, pp 37–51

Chua N-H, Blomberg F (1979) Immunochemical studies of thylakoid membrane polypeptides from spinach and chlamydomonas reinhardti. J Biol Chem 254:215–223

Clarke HGM, Freeman T (1968) Quantitative immunoelectrophoresis of human serum proteins. Clin Sci 35:403–423

Converse CA, Papermaster DS (1975) Membrane protein analysis by two-dimensional immunoelectrophoresis. Science 189:469–472

Dodge JT, Mitchel C, Hanahan DJ (1963) The preparation and chemical characteristics of hemoglobin-free ghosts of human erythrocytes. Arch Biochem Biophys 100: 119–130

Gardner E, Rosenberg LT (1969) Double diffusion in agarose: Precipitin lines which are not the result of antige-antibody reactions. Immunology 17:71–76

Goenne A, Ernst R (1978) The solubilization of membrane proteins by novel zwitterionic surfactants, Sulfobetaines. Anal Biochem 87:28–38

Grabar P, Burtin P (1964) Immunoelectrophoretic analysis. Elsevier, Amsterdam

Helenius A, Simons K (1975) Solubilization of membranes by detergents. Biochim Biophys Acta 415:29–79

Helenius A, Simons K (1977) Charge-shift electrophoresis. Simple method for distinguishing between amphiphilic and hydrophilic proteins. Proc Natl Acad Sci USA 74:529–432

Howe C, Lee LT (1969) Immunochemical study of hemoglobin-free human erythrocyte membranes. J Immunol 102:573–592

Kindmark C-O, Thorell JI (1972) Quantitative determination of individual serum proteins by radio-electroimmuno assay and use of ^{125}I-labelled antibodies. Scand J Clin Lab Invest 29: Suppl 124, 49–53

Krøll J (1976) Immunoelectrophoretic quantitation of trace proteins. J Immunol Methods 13:333–339

Laurell C-B (1965) Antigen-antibody crossed electrophoresis. Anal Biochem 10:358

Laurell C-B (1966) Quantitative estimation of proteins by electrophoresis in agarose gel containing antibodies. Anal Biochem 15:45–51

Laurell C-B (ed) (1972) Electrophoretic and electroimmuno-chemical analysis of proteins. Scand J Clin Lab Invest 29: Suppl 124

Norén O, Sjöström H (1979) Fluorography of tritium-labelled proteins in innumoelectrophoresis. J Biochem Biophys Methods 1:59–64

Owen P, Smyth CJ (1976) Enzyme analysis by quantitative immunoelectrophoresis. In: Salton MRJ (ed) Immunochemistry of enzymes and their antibodies. Wiley and Sons, New York, pp 147–202

Plesner TC (1978) Lymphocyte associated β_2-microglobulin studied by crossed radioimmunoelectrophoresis. Scand J Immunol 8:363–367

Ramlau J, Bjerrum OJ (1977) Labelling immunoelectrophoresis. A general method for increasing the sensitivity of rocket immunoelectrophoresis with ^{125}I-labelled antibodies. Scand J Immunol 6:868–871

Tanford C, Reynolds J (1976) Characterization of membrane proteins in detergent solutions. Biochim Biophys Acta 457:133–170

Svendsen PJ (1973) Fused rocket immunoelectrophoresis. Scand J Immunol 2: Suppl 1: 69

Svendsen PJ (1979) In: Deyl Z (ed) Electrophoresis, a survey of techniques and applications. pp 133–153

Svendsen PJ, Axelsen NH (1972) A modified antigen-antibody crossed electrophoresis characterizing the specificity and titre of human precipitins against Candida albicans. J Immunol Methods 1:169

Verbruggen R (1975) Quantitative immunoelectrophoretic methods: A literature survey. Clin Chem 21:5–43

Two-Dimensional Thin Layer Chromatography. Separation of Lipids Extracted from Native and Phospholipase A$_2$ Treated Human Erythrocyte Ghosts

B. ROELOFSEN and P. OTT

I. Introduction

Qualitative and quantitative analyses of (phospho)lipids play an important role in studies on the structure and function of biological membranes. Even the investigator who is primarily interested in the protein fraction of a biological membrane cannot avoid its lipid constituents. This is not only because these compounds are believed to form the structural backbone of the membranes, but also because they are involved directly in the (catalytic) function of some membraneous proteins.

It is necessary for such analyses to have at one's disposal an adequate technique for the separation of (complex) natural lipid mixtures into their individual classes. At present, Thin Layer Chromatography (TLC) can be considered as one of the most effective and versatile techniques for such separations. TLC unites qualities such as: simplicity, rapidity, ease of manipulation, sensitivity, selectivity and, last but not least, a high resolving power. However, despite this resolving power, natural lipid mixtures are usually so complex that developing the chromatogram in one dimension only is not sufficient, regardless of the variations in composition of the developing system used. Although some improvement may be achieved by multiple development with different solvents in one dimension, optimal results are only obtained by two-dimensional development, which increases the resolving power tremendously. All known two-dimensional TLC systems used for the separation of phospholipids differ mainly from one another in (minor) variations in the composition of the developing systems employed.

To demonstrate the possibilities of two-dimensional TLC, the system described by Broekhuyse (Broekhuyse 1969) has been selected to separate total lipid extracts from native as well as from phospholipase A$_2$ (PLA$_2$) treated human erythrocyte ghosts. It will be shown that a complete separation into the individual P-lipid classes can be achieved not only in the case of lipids extracted from the native membrane, but also when one is dealing with the much more complex mixture extracted from the PLA$_2$ treated ghosts. The latter fractions also contain the free fatty acids and corresponding 1-acyl lyso-derivatives produced from the 1,2-diacyl glycero-P-lipids by the hydrolytic action of the PLA$_2$.

II. Equipment and Chemicals

A. Equipment

high speed centrifuge (Sorvall RC-2B with SS-34 rotor, or equivalent) with
    tubes having a capacity of about 40 ml
clinical centrifuge with glass tubes (about 20 ml)
thermostated waterbath (37°C)
magnetic stirrer; Teflon-coated stirring bars (1.5 cm)
rotary evaporator
pear-shaped flasks (50 ml) with glass stoppers
TLC-applicator (preferable Q-fit)
glass plates (20 × 20 cm and 5 × 20 cm)
rack for transport and storage of TLC-plates
TLC-tanks (all glass) for development of chromatograms
hairdryer
oven (130°C)
spray bottles
pipettes (10, 25, 100, and 200 μl; 1.0 and 5.0 ml)

B. Chemicals

suspension of unsealed erythrocyte ghosts (human), containing 5–7 mg
    protein per ml
phospholipase A_2 (purified from either pancreas or snake venom); a solu-
    tion of 1 I.U. per 10 or 25 μl is most appropriate
5 mM $CaCl_2$, 100 mM Tris. HCl, pH 7.4
50 mM EDTA or EGTA
chloroform (p.a.)
methanol (p.a.)
ethanol (p.a.)
acetone (Techn. quality)
acetic acid, glacial (p.a.)
ammonia solution, 25% (p.a.)
silicagel ("Kieselgel 60 HR reinst", Merck, Darmstadt)
magnesium silicate ("Florisil", Merck, Darmstadt)
solid iodine
ninhydrin reagent (1 g ninhydrin in 100 ml acetone)
molybdenum blue (Zinzadze) reagent, prepared as follows (Dittmer and
    Lester 1964):

Solution I: Boil 40.1 g MoO_3 in 1 liter of 25 N H_2SO_4 until dissolved.

Solution II: 1.78 g of powdered molybdenum is added to 500 ml of solution I. Boil gently for 15 min, cool and decant any residue. Mix equal volumes of solutions I and II and dilute with 2 volumes of water. Final solution should be greenish-yellow.

III. Experimental Procedures

A. Treatment of Ghosts with Phospholipase A_2

Each of two 1.0 ml samples of a suspension of human erythrocyte ghosts (about 5–7 mg protein per ml), in a 40 ml (high speed) centrifuge tube, is diluted with 1.0 ml 100 mM Tris. HCl, pH 7.4 containing 5 mM $CaCl_2$. (Ca^{2+} is the essential cofactor for the PLA_2.) One international unit (I.U.) of PLA_2 is added to one of the two samples; the other one serves as a blank. Both samples are incubated at 37°C under continuous shaking or stirring. After 5 min the enzyme reaction is terminated by the addition of 0.2 ml 50 mM EDTA[1]. In order to remove most of the Tris and EDTA, both centrifuge tubes are filled with distilled water and the ghosts subsequently collected by centrifugation for 20 min at about 30,000 g. After (careful) removal of the supernatant, the ghosts are resuspended in 1.0 ml of distilled water.

B. Lipid Extraction (Reed et al. 1960)

The resuspended ghosts (1.0 ml) are transferred into 20 ml glass centrifuge tubes and after addition of 5 ml methanol, the mixture is stirred for 5 min. Subsequently 5 ml of chloroform is added, followed by another 5 min of stirring. After centrifugation for 10 min at 3000 g, the supernatant is collected and the extraction procedure is repeated twice. The pooled extracts are collected in a pear-shaped flask and dried under reduced pressure at 37°C. A repeated evaporation with absolute ethanol may be necessary to obtain a water-free residue. Finally, the dry residue is dissolved in about 250 μl chloroform-methanol (1:1, v/v).

---

1   A longer incubation time will lead to a complete degradation of substrates. Intentionally, conditions are chosen which result in the simultaneous presence of the original undisgested compounds and their split products; this better illustrating the high resolving power of the TLC system used

C. Preparation of TLC Plates

A series of five 20 × 20 cm glass plates are arranged on the mounting board
of the application system, with one 5 × 20 cm plate at each end of the
series, so as to facilitate an even covering of all of the five 20 × 20 cm plates.
The last traces of fat and grease (finger prints!) are removed by thoroughly
cleaning the plates with a tissue moistened with acetone. A uniform slurry
of 60 g of silicic acid (Kieselgel 60 HR reinst, Merck) plus 4.8 g [2] Mg-silicate
("Florisil", Merck) in 135 ml distilled water is prepared in an Erlenmeyer
flask (at least 300 ml). As soon as the silicic acid slurry is poured into the
applicator, the glass plates are coated with a layer of 0.75 mm thickness.
The plates are left at room temperature for a couple of hours. They are
placed in a drying rack after most of the water has been evaporated, and
activated by heating at 130°C for at least 2 h or over-night.

D. Preparation of Chromatography Tanks

To aid in saturating the chromatography tanks with solvent vapor, which is
essential to obtain optimal separation, the tanks are lined with filter paper.
As some time is needed for equilibration, the tanks should be filled with the
solvents at least *one hour before use*. Equilibration can be further facilitated
by pouring the solvent (150 ml per tank) along the filter paper-covered walls.
The tanks should be protected against any draught. The following mixtures
are used for development of the chromatogram in the first (solvent I) and
second (solvent II) dimension, respectively.

Solvent I: chloroform-methanol-conc. ammonia-H_2O
                    90   :   54   :   5.5   :   5.5 (v/v)

Solvent II: chloroform-methanol-acetic acid-H_2O
                    90   :   40   :   12   :   2   (v/v)

E. Application of Samples to the Plates

Remove the plates from the oven where they have been activated and spot
the samples as soon as the plates have been cooled to room temperature.

---

2 Broekhuyse (1969) prescribed the addition of 2% (w/w) of Mg-silicate, but in our
   experience, increasing this amount up to 6%–8% (w/w) improves the separation
   between PS and PI without any appreciable effect on the relative positions of the
   other P-lipid compounds

About half of the volume ($\sim$ 100 μl) of the samples prepared under "B" is spotted on the right hand bottom corner of the plate, at distances of about 3 cm from each edge. The plate is labeled (indicating the sample that has been spotted) at the upper left hand corner.

F. Chromatogram Development

As soon as the samples have been applied, the plates are placed in the tank containing solvent I. Care is taken that the silicic acid-coated sides are facing each other and that there is no direct contact between the silicic acid layer and the filter paper along the walls. The plates are removed from the tank when the solvent has ascended within about 2 cm of the top of the plates (about 60 min).

Before development in the second solvent (II), the first solvent has to be completely removed. This is done by means of a stream of warm air (hair dryer). This will take at least 15 min, but should be continued for a longer period if the odor of ammonia is observed. It should be stressed that incomplete removal of the first solvent will inevitably lead to a bad final result.

For development in the second direction, the plates are turned clockwise through 90° and placed into the tank containing solvent II. Finally, the plates are dried again in a stream of warm air.

G. Detection of Lipids

The following staining techniques can be used one after the other when applied in the sequence indicated.

1. Iodine Vapor. The plate is placed in a closed container saturated with iodine vapor for a few minutes; yellow or brown spots will appear, and all lipid compounds will be colored, including free fatty acids and cholesterol (see Fig. 1). The position of the various compounds is marked by circling the spots with an injection needle. The iodine is removed by heating the plate in an oven.

2. Ninhydrin Stain. The TLC plate is sprayed with ninhydrin solution (1 g ninhydrin in 100 ml acetone) and warmed in an oven or with the aid of a hair dryer. Purple spots will appear which correspond to the *amino-containing* P-lipids (PE and PS and their respective lysocompounds).

3. Phosphate Stain. The plate is sprayed with molybdenum blue (Zinzadze) reagent. *Phospho*lipids appear as blue spots within a few minutes without heating.

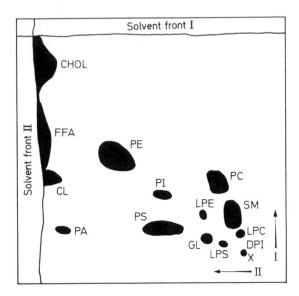

Fig. 1. Two-dimensional thin-layer chromatogram of lipids extracted from human erythrocyte ghosts which had been briefly exposed to treatment with phospholipase A₂. Abbreviations: *CHOL* cholesterol; *FFA* free fatty acids; *CL* cardiolipin (= diphosphatidylglycerol); *PA* phosphatitic acid; *(L)PE* (lyso)-phosphatidylethanolamine; *(L)PC* (lyso)-phosphatidylcholine; *(L)PS* (lyso)-phosphatidylserine; *PI* phosphatidylinositol; *SM* sphingomyelin; *GL* glycolipids; *DPI* phosphatidylinositol-phosphate; *X* origin where sample has been spotted. *Arrows* indicate direction of development with solvents I and II, respectively

IV. Comments

As can be seen from Fig. 1, a complete separation of P-lipids can be achieved. This is the case not only for the P-lipids which are present in the native red cell membranes, but even for the lyso-derivatives which may be present. Note that none of the lyso-compounds interfere with each other or any of the original compounds. It is obvious that this TLC system provides a valuable tool for studies on systems which have endogeneous PLA₂ activities or in which P-lipases are used as tools to get information on the structure and/or function of P-lipids in membranes.

Due to its resolving power, the 2D-TLC system can easily be used for reliable quantitative analyses on the composition of complex P-lipid mixtures. After staining the spots with iodine, the individual compounds can be scraped from the plate and quantitatively transferred into test tubes. After complete digestion of the organic material, for instance with 70% HClO₄ at 180°–200°C, the inorganic phosphate produced can be quantitatively determined according to various standard procedures by measuring the optical density of the blue-colored solutions obtained after reduction of the complex formed between the phosphate and molybdenate.

References

Broekhuyse RM (1969) Quantitative two-dimensional thin-layer chromatography of blood phospholipids. Clin Chim Acta 23:457–461

Dittmer JC, Lester RL (1964) A simple, specific spray for the detection of phospholipids on thin-layer chromatograms. J Lipid Res 5: 126–127

Reed CF, Swisher SN, Marinetti GV, Ede EG (1960) Studies of the lipids of the erythrocyte. I. Quantitative analysis of the lipids of normal human red blood cells. J Lab Clin Med 56: 281–289

Separation of Fatty Acids from Plant Membrane Lipids by a Combined GLC-TLC Procedure

W. EICHENBERGER

I. Introduction and Aims

Membrane properties are greatly influenced by the acyl pattern of membrane lipids. The separation and identification of fatty acids is therefore an important part of the characterization of membrane constituents. Fatty acid mixtures from plant lipids are often quite complex due to the large number of unsaturated components present in these lipids. The scope of the experiment is the separation of fatty acids on the basis of both chain length and unsaturation, and the isolation of single components in a micropreparative scale. The procedure comprises isolation of total fatty acids, methylation, and separation by preparative GLC and TLC.

II. Equipment and Solutions

A. Equipment

gas liquid chromatograph Perkin Elmer 990 (used for preparative separation), equipped with FID and stream splitter 1:50
steel column 1.8 m, 6 mm I.D., packed with DC-200 silicone 15% on Chromosorb W
gas liquid chromatograph Perkin Elmer Sigma 3 (used for analytical separation), equipped with FID. Glass capillary column 20 m, 0.25 mm I.D., coated with Carbowax 20 M
centrifuge 4 × 15 ml
UV lamp 366 nm
Rotavap
water bath
chromatography tube 1.2 cm I.D. by 12 cm long
centrifuge tubes 15 ml
conic reagent tubes 1.6 × 6 cm

pipettes 1.0, 5.0, and 10 ml
Pasteur pipettes
microsyringes 10 and 50 μl
TLC equipment
round bottom flask 100 ml

B. Solutions and Chemicals

rhodamin 6 G 0.05% in ethanol 95%
HCl – H_2O 1:1 (v/v)
diazomethane in ethyl ether (preparation see Kates, p. 526)
KOH – H_2O – ethanol 1:2:20 (w/w/v)
hexane – ethyl ether 1:1 (v/v)
hexane – ethyl ether 100:15 (v/v)
hexane – ethyl ether – acetic acid 80:18:2 (v/v/v)

NaCl saturated
silicagel G (Merck), Type 60 for TLC
silicagel 60 (Merck) 70 – 230 mesh
$AgNO_3$
nitrosomethylurea
14:0, 16:0, 18:0, 18:1, 18:2, 18:3 fatty acid methyl esters

acetone
acetic acid
dichloromethane
ethanol
ethyl ether
methanol
hexane
Pd catalyst for hydrogenation
aluminum oxide (Merck), alkaline
KOH

III. Experimental Procedure

A. Isolation and Methylation of Total Fatty Acids

Dry and powdered plant leaf material (0.8 g) is equally distributed among
four centrifuge tubes (15 ml). Each portion is extracted three times with
5 ml MeOH and once with 5 ml of ethyl ether. The extracts are collected in

a round bottom flask (100 ml) and evaporated on a Rotavap. The residue is dissolved in 1 ml of ethyl ether and the solution transferred to a centrifuge tube. The solvent is evaporated in a stream of nitrogen. To the residue 1 ml of a mixture of KOH-H_2O-ethanol 1:2:20 is added and the tube stoppered by a glass bead. The mixture is held at 70°C for 20 min. After adding 1 ml of H_2O extract three times with 5 ml of hexane-ethyl ether 1:1. The organic phase is discarded. The watery phase is acidified with 0.3 ml of HCl 1:1 and extracted again three times with 5 ml of hexane-ethyl ether 1:1. Extracts are collected in a round bottom flask and the solvent is evaporated. The residue is dissolved in 1 ml of ethyl ether and transferred to a centrifuge tube and washed with 2 ml of a saturated NaCl solution. The organic phase is transferred to a centrifuge tube and the solvent evaporated. The residue is dissolved in 0.1 ml methanol and a solution of diazomethane in ethyl ether is added in portions of 1 ml until the nitrogen evolution stops. The solvent is evaporated in a stream of N_2.

B. Purification of Fatty Acid Methyl Esters

In order to remove colored impurities (from pigments) fatty acid esters are dissolved in 0.5 ml of hexane-ethyl ether 100:15 and layered on top of a 1.2 X 2 cm column of aluminum oxide (Merck), alkaline, and eluted with 5 ml of the same solvent. The eluate is collected in a centrifuge tube and the solvent evaporated with nitrogen.

C. Separation by GLC on the Basis of Chain Length

A gas chromatograph Perkin Elmer 990 equipped with FID and a stream splitter 1:50 is used. Column DC-200 silicone 15% on Chromosorb W. Column temperature 200°C, flow 75 ml N_2/min, injector and detector temperature 230°C. Fatty acid methyl esters are dissolved in 0.5 ml of dichloromethane and 50 μl of the solution are injected. The fractions are collected in glass tubes 7 X 70 mm plugged with cotton and Silicagel 60 (70–230 mesh), which are connected to the outlet. The fraction containing the C_{16} fatty acids appears after about 8 min, the C_{18} fraction after about 16 min. Another two to three portions of 50 μl are injected and the fractions are collected in the same traps. Finally, the glass tubes are rinsed with 5 ml ethyl ether and the solution is collected in a centrifuge tube.   ˙

D. Separation on AgNO₃/Silicagel G Plates on the Basis of Unsaturation

TLC plates 20 X 20 cm are used. Five plates are obtained from 36 g Silica-gel G and 1.8 g AgNO₃ dissolved in 80 ml H₂O, activated for 90 min at 120°C and cooled in the dark. The greater part of both the C_{16} and the C_{18} fraction is spotted on a 5 cm start line. On both sides of the samples, about 50 μl of a reference mixture containing 18:0, 18:1, 18:2, 18:3 methyl esters are spotted. The plate is developed twice in hexane-ethyl ether-acetic acid 80:18:2 (v/v/v) in the dark. Spots are visualized by spraying with Rhodamin 6 G 0.05% in ethanol 95% under UV light (366 nm). The number of double bonds may be estimated by the position of spots on the TLC plate relative to standard C_{18} mono-, di-, and trienes. The spots are scraped off and extracted with 2 ml of ethyl ether in a centrifuge tube. Isolated fatty acid methyl esters may be used for further identification (e.g., by mass spectrometry).

E. Analytical Separation by GLC

A wall coated open tubular column (WCOT) loaded with Carbowax 20 M is used. Length 20 m, inner diameter 0.25 mm. Column temperature 185°C, injector and detector temperature 230°C, flow 0.5–1 ml H₂/min. These conditions may be used for analytical separation of total mixtures as well as for purity tests for single components.

The chain length of single fatty acids may be verified by comparison of their retention time after hydrogenation relative to standard saturated acids. For hydrogenation to a portion of an isolated component approx. 1 mg of Pd catalyst is added in a conical glass tube and washed down with 1 ml of acetone. Hydrogenation is obtained by a slow stream of H₂ (1 bubble per s) for 60 min at room temperature.

Reference

Kates M (1972) Techniques of lipidology. In: Work TS, Work E (eds) Laboratory techniques in biochemistry and molecular biology. North-Holland Publ Co, Amsterdam London

Isolation of Proteins

Affinity-Chromatography of Mouse Histocompatibility Antigens on Lens culinaris Lectin-Sepharose

K.J. CLEMETSON and M.-L. ZAHNO

I. Introduction – Lectins

A. Lectins are proteins or glycoproteins which bind certain sugars or oligo-saccharides specifically (Sharon and Lis 1972; Goldstein and Hayes 1978). Originally found principally in plants (Leguminosae are particularly good sources), it is known that they are widespread also in animals and recent studies have indicated their presence on the membranes of several types of human cells.

Although the commoner commercially available lectins, such as concanavalin A are often described in terms of their affinity for simple sugars – in this case glucose and mannose – in general they show higher affinities for more complex structures – oligosaccharides – containing several sugar moieties. This property can be used to pick out certain glycoproteins from a complex mixture and is extremely useful as a general approach to the purification of membrane proteins, many of which are glycosylated (Hayman and Crumpton 1972; Findlay 1974; Adair and Kornfeld 1974).

B. Coupling of Lectins to Solid Supports

Although various methods now exist for coupling proteins to solid supports while retaining biological activity, activation of beaded agarose gels with cyanogen bromide remains a popular method. It is simple and does not require complicated apparatus. The cyanogen bromide reacts with two neighboring hydroxyl groups within the agarose structure to produce both cyclic and acyclic imidocarbonates which can then react with primary amino groups in the protein to be coupled.

$$
\begin{array}{l}
-OH \\
-OH
\end{array}
+ CNBr \quad
\begin{array}{l}
-O \\
-O
\end{array}\!\!\!\!\!>\!C=NH
\; + H_2NR \qquad
\begin{array}{l}
\overset{NH}{\underset{\parallel}{}} \\
-O-C-NHR \\
-OH \\
-OH \\
-O-C-NR \\
\underset{O}{\parallel}
\end{array}
\qquad or
$$

(for discussion of the chemistry see Joustra and Axén 1975)

After coupling it is necessary to wash the protein-agarose thoroughly in order that unreacted protein or protein held by labile linkages is removed.

When lectins are being coupled it is advisable to add the specific sugar for the lectin to the lectin/coupling buffer in order to protect the binding site.

C. Affinity-Chromatography on Insolubilized Lectins

This is now a well-established and popular technique either for purifying glycoproteins or polysaccharides or for removing these from protein preparations and many commercial preparations of lectins and insolubilized lectins can now be bought. Where preparative applications are envisaged rather than preliminary investigations, it is generally worth taking the trouble to make the lectin absorbents, as many lectins are easy to prepare and purify in large quantities.

When integral membrane components are to be separated by any form of affinity-chromatography the first requirement is that they be solubilized and brought into a monomeric state so that when they bind to the affinity-absorbent, nonbound material can be eluted. This solubilization is generally carried out with types of detergent which do not affect the biological activity of either the membrane component or of the affinity-absorbent.

D. Histocompatibility Antigens

The cells of warm-blooded animals carry to a greater or lesser extent a range of antigens which are characteristic of the individual animal and which provide a mechanism for recognizing self from foreign cells. These — histocompatibility — antigens expressed on kidney or heart transplants are responsible for their rejection by the recipient unless the donor is genetically identical (i.e., an identical twin) or the host's immune response is suppressed. These complications can be minimized by careful matching of donor with host histocompatibility antigens. The normal biological function of these antigens is not yet clear; amino-acid sequencing studies have shown that there exist considerable homologies between human histocompatibility antigens and immunoglobulin G heavy chains. It is thought that they may play a role in the recognition of virally transformed or neoplastic cells by the immune system.

These histocompatibility antigens are glycoproteins which bind to *Lens culinaris* lectin and can be eluted with methyl-α-D-mannopyranoside (Snary et al. 1974; Clemetson et al. 1976b). This provides a simple method for purifying the antigens.

E. Aims of the Experiment

The object of the experiment is to demonstrate:

1. Purification of a lectin by affinity chromatography.
2. A cell agglutination assay for lectin activity.
3. Coupling of this lectin to agarose to prepare insolubilized lectin.
4. Solubilization of mouse tumor cell membranes. Use of the lectin-agarose packed in a column to isolate histocompatibility antigens from these solubilized membranes.

II. Equipment and Materials

A. Apparatus

2 fraction collectors
spectrophotometer (or, better, UV monitors)
fume hood
ultracentrifuge
SW 50.1 Rotor or equivalent
sintered glass funnel D3, 6.5 cm diameter
1–2 liter Buchner funnel
magnetic stirrer
column to pack 20 ml coupled lectin-Sepharose (12 × 1.5 cm)
column to pack 100 ml Sephadex G-150 (20 × 2.5 cm)

B. Materials

Lens culinaris lectin ∼ 40 mg prepared by the method of Sage and Green (1972)
partially purified *Lens culinaris* lectin (method of Sage and Green 1972, up to and including DEAE column)
P-815 mouse mastocytoma membranes, 2 ml 10–12 mg protein/ml in 0.005 M Tris/HCl, 0.15 M NaCl pH 7.4 prepared according to Clemetson et al. (1976a)
P-815 cells – grown in in vitro cell culture

Membranes from other cells containing histocompatibility antigens may be used such as murine lymphocytes, human lymphocytes or other tumor lines grown in culture.
Chemicals used are mostly Merck P.A. unless otherwise noted.

III. Experimental Procedures

A. Purification of Lens culinaris Lectin by Affinity-Chromatography on Sephadex G-150

A partially purified extract of *Lens culinaris* (lentils, available in food shops) is made 0.15 M in NaCl and then loaded onto a column of Sephadex G-150 (20 × 2.5 cm) which has been pre-equilibrated with 0.015 M PO$_4$ buffer, 0.15 M NaCl pH 7.5. When the optical density of the effluent drops back to that of the buffer the eluent is changed to 2% glucose in the same buffer. A sharp, high O.D. peak elutes with the glucose front (Fig. 1). The peak fractions are pooled and dialyzed versus 0.015 M PO$_4$, 0.15 M NaCl pH 7.5 to remove glucose.

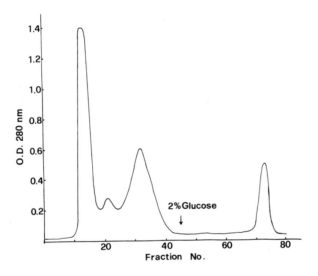

Fig. 1. Purification of *Lens culinaris* lectin by affinity chromatography on Sephadex G-150

B. Determination of Agglutination Titer of Lectin

Serial two-fold dilutions of the lectin fractions to be tested are prepared in a microtiter plate (Fig. 2).

1. To each hole is added 20 μl PBS.
2. To the first hole is added 20 μl lectin, 20 μl of the mixture is then withdrawn and added to the second hole and so on.

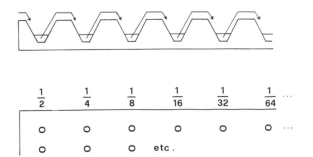

$$\frac{1}{2} \qquad \frac{1}{4} \qquad \frac{1}{8} \qquad \frac{1}{16} \qquad \frac{1}{32} \qquad \frac{1}{64} \quad \cdots$$

o   o   o   o   o   o   ...

o   o   o   etc.

Fig. 2. Preparation of two-fold serial dilutions of lectin in a microtiter plate

P-815 cells are centrifuged down from medium and washed twice with phosphate buffered saline (PBS). They are resuspended in PBS to 2.5×10^6 cells/ml. 20 μl of cell suspension is added to each lectin dilution in the microtiter plate. The plate is incubated in a 5% CO_2 atmosphere at 37°C for 30 min. The cell suspension in each hole is then removed, placed on a microscope slide, covered with a cover slip and examined under the microscope at a magnification of about 60-fold. The agglutination titer is the highest dilution at which agglutination of the cells is perceptible. This assay may also be carried out using human or rabbit erythrocytes as described by Sage and Green (1972).

C. Activation of Sepharose and Coupling of Lectin

1. Sepharose 4B (20 ml) is washed thoroughly with at least 1 liter of distilled water on a sintered glass funnel with gentle suction from a water pump and suspended in 30 ml of $2 M \ Na_2CO_3$. To this slowly stirred suspension is quickly added from a pipette 1 ml of a solution of CNBr[1] in acetonitrile (0.5 g/ml). The stirring is changed to rapid for 1.5—2 min, then the gel is quickly poured into a sintered glass funnel and washed as quickly as possible with 200 ml each of $0.1 M$ NaHCO$_3$, H_2O and $0.2 M$ NaHCO$_3$ in succession (March et al. 1974). The activated gel is immediately added to 20 ml of a solution containing 1—2 mg/ml *Lens culinaris* lectin (exact amount depends on the individual lectin preparation), 2% methyl-α-D-glucopyranoside and $0.1 M$ NaHCO$_3$ (check pH of solution — it should be ~ 8.5) precooled to 4°C. The suspension is stirred slowly at 4°C for 20—24 h and is then washed with:

---

1   As CNBr is highly lachrymogenic and also both very toxic and carcinogenic it should only be used in a fume hood, should be weighed in a stoppered tube, and should be handled only while wearing gloves

2. (a) 200 ml 0.1 M sodium acetate buffer pH 4 (with acetic acid) in 0.5 M NaCl.
   (b) 200 ml 2 M urea in 0.5 M NaCl.
   (c) 200 ml 0.1 M NaHCO$_3$ in 0.5 M NaCl.
   (d) 500 ml phosphate buffered saline pH 7.4 (0.01 M PO$_4$, 0.15 M NaCl).

When not in use the gel is stored in phosphate buffered saline pH 7.4 containing 0.05% NaN$_3$.
The gel is packed into an appropriately sized column and washed to remove azide before use. Where sodium deoxycholate solutions are to be used the NaCl should be washed out with distilled water.

D. Solubilization of Membranes

P-815 membranes (2 ml, 20–30 mg protein, stored at − 70°C) are thawed out and 24 mg of solid sodium deoxycholate (DOC) added. The mixture is vortexed until the DOC has dissolved and is then left for 20 min at room temperature with occasional vortexing. The solubilized membranes are then centrifuged for 1 h at 35,000 rpm in an SW50.1 rotor in a Beckman L-2 centrifuge using centrifuge tubes which have been cut to hold 2 ml. While the centrifuge is running the lectin column is washed with 50 ml of distilled water and then 50 ml of 0.5% DOC in distilled water.

E. Affinity-Chromatography

The supernatant is carefully removed from the centrifuge tube and after allowing the buffer in the column to sink to the surface of the gel is loaded onto the column. The supernatant is allowed to sink to the surface of the gel and then 0.5% DOC is added. The column is washed with this buffer and 1 ml fractions collected (approximately 35 drops) until the O.D. 280 nm of the column effluent returns to that of the buffer (approximately 20–25 fractions). The eluting buffer is then changed to 2% methyl-α-D-mannopyranoside, 0.5% DOC for 20 ml and then back to 0.5% DOC. A small peak of protein elutes and the O.D. 280 nm returns to baseline (see Fig. 3). The column should then be washed with distilled water and then with storage buffer, 0.005 M phosphate pH 7.4, 0.15 M NaCl, 0.05% NaN$_3$.

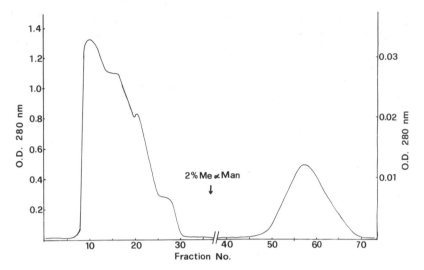

Fig. 3. Purification of mouse histocompatibility antigens by affinity chromatography on *Lens culinaris* lectin – Sepharose 4B. *Me αMan* is an abbreviation for Methyl-α-D-mannopyranoside

IV. Results

For both column chromatography experiments the optical density at 280 nm of the fractions should be plotted against the fraction number. The fraction number at which changes in eluting buffers were made should be noted on these graphs.

The relative agglutination of cells produced by each dilution of lectin should be recorded based on a scale from 3 + (all cells agglutinated in large masses) to + (some cells agglutinated in 2's and 3's) to − (no cells agglutinated). The lowest concentration of lectin for which + was recorded should be noted as the lectin titer.

V. Comments

If an antiserum is available against the histocompatibility antigens separated in this experiment and sufficient time is available, this experiment may be extended by using an inhibition of cytotoxicity assay to demonstrate which fractions contain the histocompatibility antigens (Clemetson et al. 1976a,b).

Alternatively this experiment may be combined with some of the immuno-electrophoresis experiments described earlier (O.J. Bjerrum) if suitable antisera are available.

References

Adair WL, Kornfeld S (1974) Isolation of the receptors for wheat germ agglutinin and the *Ricinus communis* lectins from human erythrocytes using affinity chromatography. J Biol Chem 249:4694−4704

Clemetson KJ, Gerber A, Bertschmann M, Lüscher EF (1976a) Solubilization of histo-compatibility and tumour-associated antigens of the P-815 murine mastocytoma cell. Eur J Cancer 12:263−270

Clemetson KJ, Bertschmann M, Widmer S, Lüscher EF (1976b) Water-soluble P-815 mastocytoma membrane antigens. Immunochemistry 13:383−388

Findlay JBC (1974) The receptor proteins for concanavalin A and *Lens culinaris* phytohemagglutinin in the membrane of the human erythrocyte. J Biol Chem 249: 4398−4403

Goldstein IJ, Hayes CE (1978) The lectins: Carbohydrate binding proteins of plants and animals. In: Tipson RS, Horton D (eds) Advances in carbohydrate chemistry and biochemistry, vol 35. Academic Press, London New York, pp 127−340

Hayman MJ, Crumpton MJ (1972) Isolation of glycoproteins from pig lymphocyte plasma membranes using *Lens culinaris* phytohaemagglutinin. Biochem Biophys Res Commun 47:923−930

Joustra M, Axén R (1975) Stability of the binding groups generated by CNBr activation of agarose. In: Peeters H (ed) Protides of biological fluids, vol 23. Pergamon Press, Oxford, pp 525−529

March SC, Parikh I, Cuatrecasas P (1974) A simplified method for cyanogen bromide activation of agarose for affinity chromatography. Anal Biochem 60:149−152

Sage HJ, Green RW (1972) Common lentil *(Lens culinaris)* phytohemagglutinin. In: Ginsburg V (ed) Methods in enzymology, vol 28. Academic Press, London New York, pp 332−339

Sharon N, Lis H (1972) Lectins: Cell-agglutinating and sugar-specific proteins. Science 177:949−959

Snary D, Goodfellow P, Hayman MJ, Bodmer WF, Crumpton MJ (1974) Subcellular separation and molecular nature of human histocompatibility antigens (HL-A). Nature (London) 247:457−461

Isolation of a Membrane Protein: The Phospholipase A₂ from Sheep Erythrocyte Membranes

P. ZAHLER, C. WÜTHRICH, and M. WOLF

I. Introduction

Red cell membranes from sheep contain a phospholipase A_2 (EC 3.1.1.4) with the following characteristic properties: marked preference for phosphatidylcholine (PC), specificity for the fatty acid at the 2-position, requirement for Ca^{2+}, alkaline pH optimum, activation by various detergents and stability against denaturing agents. The enzyme is present in ruminant erythrocytes and is thought to play a role in maintaining the very low phosphatidylcholine content of the ruminant red cell membrane. Its main function, however, seems to be the liberation of free fatty acids from PC from the plasma lipoproteins in favour of the peripheral tissue.

The isolation of the phospholipase A_2, which is present in very small amount in the membrane (less than 0.05% of the membrane proteins), is performed in presence of detergents or organic solvents using the techniques of molecular sieving and affinity chromatography. We think that it provides a good example for the problems encountered when isolating a minor hydrophobic membrane protein.

II. Flow Sheet of the Isolation

Membranes
(sheep red cell ghosts,
isolated according to
Dodge et al.)

    extracted with 1 mM   *Low salt extraction* resulting in a depletion of extrinsic membrane proteins (mostly spectrin)
    glycylclycine (pH 8)
    centrifuged

Extracted membranes   *Solubilization* of the membranes 2.5 mg protein/ml
    solubilized with 0.5%
    dodecyl sulphate,
    5% cholate

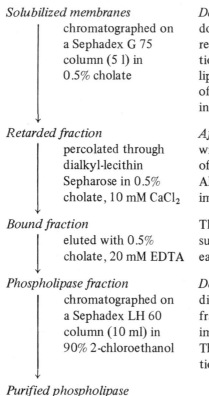

Solubilized membranes
  chromatographed on
  a Sephadex G 75
  column (5 l) in
  0.5% cholate

Detergent replacement. Gelfiltration does not only accomplish the detergent replacement but also a 60-fold purification, due to the fact that the phospholipase does not reaggregate (as most of the other membrane proteins do) in the absence of SDS

Retarded fraction
  percolated through
  dialkyl-lecithin
  Sepharose in 0.5%
  cholate, 10 mM $CaCl_2$

Affinity chromatography is performed with a nonsplitable alkyl-ether analog of lecithin which was coupled to AH-Sepharose 4B by using a carbodiimide coupling procedure

Bound fraction
  eluted with 0.5%
  cholate, 20 mM EDTA

The enzyme needs Ca^{2+} to bind the substrate so that binding and elution can easily be regulated by Ca^{2+} and EDTA

Phospholipase fraction
  chromatographed on
  a Sephadex LH 60
  column (10 ml) in
  90% 2-chloroethanol

Delipidation. After concentration, dialysis and drying, the phospholipase fraction still contains lipids and some impurities from the cholate buffer. The delipidation is done by gel filtration in 2-chloroethanol

Purified phospholipase
  (overall purification
  ∼ 2500–3000-fold)

III. Experimental Part

As the quantity of the phospholipase does not allow to follow the enzyme during its isolation by means of spectroscopic methods, one is obliged to control the isolation by an enzymatic assay.

The phospholipase A_2-assay and the last step of the isolation, consisting in a column chromatography on Sephadex LH 60 in the organic solvent 2-chloroethanol (Zahler et al. 1967)) will be described more extensively. For a complete description of the isolation procedure see Kramer et al. (1978).

A. Assay of Phospholipase A_2 Activity

1. Materials

sheep red cell ghosts (isolated according to Dodge et al. 1963)
L-α-lecithin (from egg, Koch-light, grade 1)
[^{14}C]-lecithin (from Chlorella pyrenoidosa, overall marked, NEN)
2-chloroethanol (Fluka p.a.)
Sephadex LH 60 (Pharmacia)
butyl-PBD (Ciba/Geigy)

The solvents were from Merck (reinst); all the other chemicals were reagent grade and from either Merck, Sigma or Fluka.

glas column ϕ 1 cm, 15 ml
fraction collector
bench top centrifuge (BHG)
scintillation counter (Packard Tri Carb 2450)
spectrophotometer (Unicam SP 1750)

2. Procedure

20 μl of a lecithin solution (in chloroform-methanol 1:1) corresponding to
    0.2 mg or 0.27 μmol lecithin (PC)
and 5 μl of a [^{14}C]-lecithin solution in benzol-ethanol 1:1) corresponding to
    $25 \cdot 10^{-3}$ nmol and $\sim 5 \cdot 10^4$ cpm
are evaporated in a test tube under N_2 at $37°C$.

An appropriate amount of enzyme-containing fraction ($\triangleq$ 100 ng pure enzyme) is added to the dried substrates, then 10 mM glycyl-gylcine buffer pH 8 containing 0.5% cholate, 0.2 M KCl and 0.02% Na-acid are added to a total volume of 1 ml. The reaction starts by adding 40 μl of a 0.2 M $CaCl_2$-solution.

After incubation at $37°C$ for 2 h with shaking, the reaction is stopped with 60 μl of a 0.2 M EDTA-solution. 0.2 ml of the incubation mixture are brought directly into a scintillation counting vial ($\triangleq$ total), another 0.5 ml are transferred into a second test tube, where the extraction of the released free fatty acids is performed:

add 2 ml of di-iso-propylether-methanol 99:1 and 250–300 mg of magnesium silicate,
stir vigorously on a Vortex during 30 s, leave it for 10 min and stir again.

Then the mixture is centrifuged in a BHG-centrifuge at 3000 rpm for 10 min to separate the two phases. 1 ml of the clear supernatant (organic phase) containing the fatty acids is transferred into a counting vial (take care that no magnesium silicate is picked up!).

After addition of 5 ml of methanol and 10 ml of a scintillation fluid (Toluene containing 7 g butyl-PBD per l), the radioactivity is measured in a Scintillation Counter.

3. Remarks

The experiment can conveniently be made with up to 12 different incubation mixtures.

A blank without enzyme should always be taken along (to test for the unspecific hydrolysis and possible mistakes of the extraction).

4. Results

On the assumption that only the free fatty acids are extracted in the organic phase (lysolecithin and nonsplit lecithin remaining adsorbed to magnesium silicate) and that in the overall marked lecithin the radioactivity is distributed at a rate of 1.25:1 between the two moieties lysolecithin and fatty acid (in 2-position), the calculation of "% lecithin hydrolyzed" can be done as follows:

$$\% \text{ PC hydrolyzed}/_{2\,h} = \frac{2 \times \text{cpm}_{\text{(fatty acids)}}}{\text{cpm}_{\text{(total)}}} \times 100$$

Then the blank should be deduced and the enzymatic acticity calculated in μmol PC split per hour or minute (U). Where possible, the specific activity (U per mg) should be calculated.

Each elution process is controlled by measuring the enzymatic activity. Based on these values, an elution diagram has to be drawn. Calculate the yield of the activity in the analyzed fractions.

B. Sephadex LH-Chromatography in 90% 2-Chloroethanol

This last step in the purification procedure consists in the removal of lipids and some impurities from the cholate buffer.

The freeze dried phospholipase preparation (~ 100 μg) is taken up in 90% 2-chloroethanol, shaken for 30 min at 37°C and applied carefully to a 1—12 ml glass-column containing Sephadex LH 60 equilibrated with 90%

2-chloroethanol. Take care that the solvent does not come into contact with any plastic material.

Fractions of ~ 0.6 ml are collected at a flow rate of about 4 ml per hour. The elution is followed by measuring the optical density at 280 and 220 nm.

The phospholipase A_2 should appear in a first peak in the void volume (after $\sim$ 3.6 ml), whereas a second peak is fully included ($\sim$ 10 ml) consisting of low molecular weight components (lipids, cholate, impurities). Draw the elution pattern.

It would be interesting to perform the enzymatic assay (see III. A) with some chosen fractions of the LH 60-chromatography. For this purpose 15 μl of each fraction are an appropriate amount. However, the 2-chloroethanol should be evaporated (2 h in exsiccator) after adding the 15 μl solution to the dried substrates (see III.A.1).

Alternatively, the Sephadex LH 60-chromatography experiment can also be done with a membrane preparation (ghosts) in order to delipidate the bulk of the membrane proteins. In this case, about 50 μl of ghosts ($\sim$ 10 mg protein/ml) are solubilized in 450 μl 100% 2-chloroethanol.

References

Dodge CT, Mitchell CD, Hanahan DJ (1963) Arch Biochem Biophys 100:119–130

Kramer RM, Wüthrich C, Bollier C, Allegrini P, Zahler P (1978) Biochim Biophys Acta 507:381–394

Zahler P, Wallach DFH, Lüscher EF (1967) Protides of the Biologic Fluids, vol 15, pp 69 ff

Isolation and Purification of the Nicotinic Acetylcholine Receptor from Torpedo Electric Organ

B.W. FULPIUS, N.A. BERSINGER, R.W. JAMES, and B. SCHWENDIMANN

I. Introduction and Aims

The nicotinic acetylcholine receptor (nAcChR) can be extracted and purified to homogeneity from fish electroplaques. These organs are regarded as modified skeletal muscles with a very rich innervation of the cholinergic type. Those from the electric ray *Torpedo* are the richest known source of nAcChR. They contain 12,000–15,000 receptors per μm^2 on the postsynaptic membrane.

The extraction of the receptor requires the use of a nonionic detergent throughout the isolation and purification procedures in order to maintain it as a soluble complex, active toward nicotinic cholinergic ligands. The purification rests mainly on the use of affinity chromatography. This technique requires a specific cholinergic ligand with favorable dissociating properties (i.e., cobra α-neurotoxins) covalently coupled to a solid matrix (i.e., Sepharose 4B) in order to adsorb nAcChR selectively.

The chromatography results in the retention, on the matrix, of nAcChR and elution of the macromolecules devoid of affinity for the ligand. The receptor is recovered by displacing it from the solid support with a solution containing a ligand of similar cholinergic specificity (i.e., hexamethonium). This ligand is dialyzed or washed out in a subsequent step. Affinity chromatrography, even repeated twice, does not yield pure nAcChR unless it is followed by centrifugation in a sucrose gradient, chromatography by anion-exchange or, as described here, chromatography on hydroxylapatite.

The pure receptor is a glycoprotein which has a specific binding activity for post-synaptic α-toxins of about 10,000 nmol/g protein. When sedimented on sucrose density gradients, it appears under two forms, a monomer and a dimer, with sedimentation coefficients of about 9s and 13s. Molecular weights of 250,000 and 500,000 daltons have been calculated for the monomer and dimer respectively. The receptor is composed of four subunits: α (42,000), β (49,000), γ (57,000) and δ (65,000) in a molar stoichiometry of 2:1:1:1. This is consistent with the molecular weight of 250,000 daltons attributed to the 9s form. There are two α-neurotoxin binding sites per monomeric molecule and these are located on the α-subunits. (For a review, see Heidmann and Changeux 1978.)

II. Equipment, Chemicals and Solutions

A. Equipment

1. Access to

a cold room (+ 4°C)
a refrigerated centrifuge (up to 26,000 g) with a rotor of 200 ml capacity
    (e.g., Sorvall RC 2B or 5B, rotor SS34)
a refrigerated ultracentrifuge (up to 100,000 g) with a rotor of 200 ml
    capacity (e.g., Beckman Spinco, rotor Ti 60)
a gamma counter
a spectrophotometer allowing measurements at 750 nm
a fraction collector with a peristaltic pump, an UV detector (280 nm) and
    a recorder
a high speed homogenizer with a 200 ml container (e.g., VirTis "45")

2. On the Bench:

a magnetic stirrer with Teflon-coated magnetic rods
a dissection board and a knife
glassware, automatic pipettes (5 to 200 μl) and disposable plastic tubes
    (2.5 and 10 ml capacity)
ultrafilters (0.22 μm pore size, maximum diameter 2.5 cm), filter holders
    and plastic syringes (5 ml)
dialysis tubing (about 1 X 50 cm) and sterile plastic tubes (5 ml capacity)
DEAE filter discs (Whatman DE 81, diameter 2.3 cm) and a filter holder
    (e.g., Millipore XX 1002502) with a funnel (e.g., Millipore XX 1002514)
    connected to a vacuum flask (about 1 liter)
plastic gloves
a glass or plastic chromatography column (2 X 25 cm) for the affitiny gel
a glass or plastic chromatography column (1 X 10 cm) for the hydroxylapa-
    tite gel
plastic tubing (Intramedic diameter about 1 mm) and connections for
    column chromatography

B. Chemicals

frozen electric organ from *Torpedo marmorata* are supplied by the station
    of Marine Biology, Arcachon, France

50 ml bed volume of α-cobratoxin substituted Sepharose 4B (total capacity about 500 nmol) prepared according to Klett et al. (1973)

10 ml hydroxylapatite suspension (Spheroidal hydroxylapatite from BDH or Biogel HTP from BioRad Laboratories)

a few grams of Sephadex G-200

Triton X-100 (GLC-Grade from Merck)

[^{125}I]-α-bungarotoxin (MIT) approximately 1 μM in Na-phosphate (0.15 M, pH 6.5) containing BSA [0.05% (w/v)] prepared according to James et al. (1980)

hexamethonium chloride recrystallized from hot ethanol (about 10 g)

C. Solutions

The following solutions should be prepared in advance, kept at 4°C and contain sodium azide, 0.003% (w/v)

Buffer I (0.5 l):   Tris-HCl (0.05 M, pH 7.4), NaCl (0.1 M), EDTA (0.001 M)
Buffer IIA (2 l):   Buffer I containing Triton X-100 [1% (v/v)]
Buffer IIB (0.5 l):   Buffer IIA containing NaCl (1.0 M)
Buffer IIC (0.2 l):   Buffer IIA containing hexamethonium chloride (0.05 M)
Buffer IIIA (0.2 l):   Tris-HCl (0.01 M, pH 7.4), NaCl (0.1 M), Triton X-100 [0.05% (v/v)], EDTA (0.001 M)
Buffer IIIB (0.2 l):   Buffer IIIA containing K-phosphate (0.5 M, pH 7.4)
Buffer IV (2 l):   Buffer I containing Triton X-100 [0.05% (v/v)]
Buffer V (2 l):   NaCl (0.01 M), Tris-HCl (0.01 M, pH 7.4)

solutions for protein determination according to Lowry et al. (1951) modified by Dulley and Grieve (1975) in order to eliminate interference by Triton X-100

III. Experimental Procedures

The original description of the homogeneization, extraction and purification of nAcChR was published by Klett et al. (1973), including the method for assaying receptor activity.

Day 1

Organ homogeneization and nAcChR solubilization

8 am[1]   A frozen electric organ is cleaned from skin and connective tissue and cut into small cubes (about 1 X 1 X 1 cm, 100 g total). These cubes are homogenized (75% of the maximal speed) in 100 ml ice-cold buffer I for 90 s.

9.00   One starts to wash, at room temperature, the affinity chromatography column with buffer IIA. The homogenate (200 ml total) is sedimented (26,000 g; 4°C; 15 min). The supernatnat is discarded and the pellet (about 30 g) resuspended in an equivalent volume of ice-cold buffer I by moderate homogeneization (20% of the maximal speed for 15 s).

9.45   Triton X-100 is added dropwise, whilst stirring, to a final concentration of 1% v/v. The resulting mixture is then thouroughly stirred for 90 min in the cold room.

11.15   The mixture is centrifuged (100,000 g; 4°C; 60 min).

12.30   The supernatant is recovered as *the crude extract.*

Assay for nAcChR

Before loading the column, the crude extract is tested for activity in the following way:

20 µl crude extract are diluted with 400 µl buffer IIA in order to constitute a stock of diluted receptor which will be used for protein determination (not described) and α-bungarotoxin binding capacity (see below). Aliquots (0, 5, 10, 15, and 20 µl) of the stock solution diluted 1:2 are pipetted in 2 ml disposable plastic tubes containing 200 µl buffer IIA and 10 µl [^{125}I]-α-Bgt (equivalent to 10 pmol). Each tube is incubated for 1 h, after careful mixing, and filtered as follows: two discs of DEAE-cellulose are placed on a filter holder connected to a vacuum flask. An appropriate funnel is clamped to the holder and filled with 10 ml buffer V. The system is connected to a vacuum line and, after a steady rate of filtration has been established, the solution to be analyzed is poured into the funnel and filtered (rinse the tubes containing the samples twice and don't let the discs dry). The discs are subsequently rinsed with buffer V (10 ml), dried and transferred to the appropriate vials for counting in the gamma counter. Each set of filters is counted for 1 min and the specific activity is expressed as pmol α-bungarotoxin bound per mg protein.

2 pm   *Affinity chromatography* (the column is equilibrated in buffer IIA). The expected amount of nAcChR (100–200 nmol α-bungarotoxin

---

1   The time schedule is tentative

binding sites) in the whole crude extract is loaded on the affinity column, one third of bed volume at a time with 15 min incubation after each load. 15 min after the last load, the column is washed wtih buffer IIA (spontaneous flow rate). All the material collected during loading and the first 100 ml wash is pooled for testing the receptor activity which should not exceed 30% to 40% of that contained in the total load. (Once this test is performed, discard this material which contains nAcChR with altered properties.) The column is further washed in the following sequence: buffer IIA (100 ml), buffer IIB (50 ml), buffer IIA (100 ml). It is left at 4°C overnight, all connections being checked to prevent leakages.

Day 2

Recycling through the hydroxylapatite column (the column is equilibrated in buffer IIC)

12.00   Introduce one third of bed volume of buffer IIC on the affinity column. Connect the column outlet to the top of the hydroxylapatite column and then connect the hydroxylapatite column outlet (via a peristaltic pump) to the top of the affinity column. Pump (30 ml/h) for 18 h in the cold room (see diagram in Klett et al. 1973).

Day 3

Hydroxylapatite column elution (performed at room temperature)

8.00   Connect the column outlet to the fraction collector and wash the column with 10 volumes of buffer IIIA in order to remove excess Triton X-100 (check absorbance at 280 nm).

2.00 pm Elute nAcChR with buffer IIIB and collect 1 ml fractions. The receptor elutes in a few fractions characterized by high UV absorbance. Check these fractions for toxin binding (see above) and pool those with the highest receptor content (5–10 fractions). Dialyze the pool overnight at 4°C against 2 liters of buffer IV.

Day 4

Concentration of nAcChR

8.00   Concentrate the receptor pool by covering the bag with dry Sephadex G 200. Remove delicately by hand the wet gel every 30 min until a 5- to 10-fold concentration is achieved. Carefully aspirate the bag content in a plastic syringe and ultrafilter this solution directly into a 5 ml sterile tube. The resulting nAcChR solution is then tested for protein content and toxin binding as mentioned above. A final yield of 9—10 nmol toxin binding sites per mg protein is expected.

IV. Comments

Although *Torpedo* electric organs are known to be the richest souce of nAcChR, one can still observe, from one organ to another, variations in the total receptor content due to the age of the fish and its state of conservation. Proteolytic activity is found in electric organs, as in any biological material. In this respect, basic precautions against receptor degradation have to be taken, namely work at 4°C whenever possible and working rapidly, use of inhibitors of proteolytic enzymes (e.g., EDTA). Although these precautions do not provide full protection, they seem satisfactory in the present context.

Triton X-100 has been used as nonionic detergent for solubilization because of its wide use by scientists preparing nAcChR, but other detergents (Emulphogen BC-720, Tween 80, Brij 35, NP 40, etc.) have been used successfully. Hexamethonium has been selected to elute nAcChR from the affinity chromatography since it competes effectively at the toxin binding site on the receptor (Maelicke et al. 1977). Other ligands can be used in this respect (carbamoylcholine, decamethonium), apparently without major advantages.

Recently, the use of an additional purification step based on chromatography on ConA-Sepharose has been proposed (Dolly et al. 1977). Elution is obtained with α-methylmannoside plus α-methylglucoside. This represents definitely an improvement toward the purification of an homogeneous glycoprotein population. As the receptor-specific activity is not modified by this additional step, it has not been introduced in the present protocol.

The purpose of the preparation was to provide a detergent-solubilized membrane protein for reincorporation into an artificial lipid phase. In this context, one should not forget that the receptor, once solubilized, might display properties not necessarily related to its function in situ.

References

Dolly OJ, Barnard EA, Shorr RG (1977) Characterization of acetylcholine-receptor protein from skeletal muscle. Biochem Soc Trans 5:168–170

Dulley JR, Grieve PA (1975) A simple technique for eliminating interference by detergents in the Lowry method of protein determination. Anal Biochem 64: 136–141

Heidmann T, Changeux JP (1978) Structural and functional properties of the acetylcholine receptor protein in its purified and membrane-bound states. Annu Rev Biochem 47: 317–357

James RW, Bersinger NA, Schwendimann B, Fulpius BW (1980) Characterization of iodinated derivatives of α-bungarotoxin. Hoppe-Seyler's Z Physiol Chem 361: 1517–1524

Klett RD, Fulpius BW, Cooper D, Smith M, Reich E, Possani LD (1973) The acetylcholine receptor. I. Purification and characterization of a macromolecule isolated from *Electrophorus electricus*. J Biol Chem 248:6841–6853

Lowry OM, Rosebrough NJ, Farr AL, Randall RJ (1951) Protein measurement with the Folin Phenol reagent. J Biol Chem 193: 265–275

Maelicke A, Fulpius BW, Klett RP, Reich E (1977) Acetylcholine receptor. Responses to drug binding. J Biol Chem 252:4811–4830

Preparation of Spectrin-Free Vesicles from Human Red Blood Cells

P. OTT and U. BRODBECK

I. Introduction

An important prerequisite in the study of erythrocyte membrane structure and function is the availability of suitable membrane preparations. Several methods by which such "ghosts" can be prepared have been published by various authors. A summary of some of these methods has been presented by Schwoch and Passow (1973). Intact ghosts, containing all intrinsic (intramembraneous) and extrinsic proteins, as well as cytoskeletal components, can further be converted to yield membrane fragments such as spectrin-depleted ghost or inside-out vesicles, lacking one or more membrane constituents (Steck, 1974).

Alternatively, spectrin-free vesicles may be produced by prolonged incubation (at least 20 h) of intact red blood cells in glucose-free isotonic buffer systems. These vesicles, which are thought to arise as a consequence of ATP-depletion of red blood cells, have been considered as a promising model system for the study of certain aspects of membrane architecture and function (Lutz et al. 1977). However, the possibility of artifacts arising from metabolic starvation of red blood cells and long incubation times (with possible formation of proteolytic breakdown products) should be kept in mind.

The experiment detailed below describeds the formation of spectrin-free vesicles from intact red blood cells without ATP-depletion. The maximum incubation time needed is only 6 h, which greatly reduces the possibility of artifacts.

II. Equipment and Chemicals

A. Equipment

high-speed centrifuge (Sorvall RC-2B with SS-34 rotor, or equivalent) with tubes of a capacity of about 30–40 ml
clinical centrifuge with adaptors for glass tubes of capacities from 10 to 250 ml

thermostated waterbath (30°C)

magnetic stirrer; Teflon coated stirring bars (1 cm)

sonicator (bath type sonicator or tip probe sonicator can be used, e.g., MSE
    ultrasonic disintegrator MK 2)

equipment for column chromatography including fraction collector, peris-
    taltic pump and UV flow-through monitor (Uvicord or equivalent;
    optional)

B. Chemicals

dimyristoylphosphatidylcholine = DMPC (e.g., from FLUKA, Buchs, Swit-
    zerland)

10 mM Tris-HCl, containing 144 mM NaCl, pH 7.4 = TBS (Tris-buffered
    saline)

reagents for acetylcholinesterase assay, prepared as follows: 5 mg 5,5'-
    Dithiobis-(2-nitrobenzoic acid) = DTNB and 30 mg of acetylthiocholine-
    iodide are dissolved in 100 ml sodium phosphate buffer, 100 mM, pH 7.4,
    containing 0.5% Triton X-100 (v/v)

anticoagulant: Acid Citrate Dextrose-solution = ACD, is prepared as follows:
    16.5 g glucose, 11.7 g sodium citrate and 5.44 g citric acid are dissolved
    in 500 ml of water

fresh blood sample (10 ml)

III. Experimental Procedure

A. Preparation of Blood Samples

Samples are collected from adult healthy donors and immediately and carefully added to the ACD-solution. A volume of 2.2 ml of ACD-solution is
used for 10 ml fresh blood. The erythrocytes are pelleted by centrifugation
at 3000 rpm for 15 min in a clinical centrifuge and the plasma and buffy
coat are carefully removed by aspiration. Three washing steps are carried
out using the TBS-solution. The washed and packed erythrocytes are used
in the subsequent incubation experiments.

B. Preparation of DMPC Dispersion

DMPC is dried in a test tube from a stock solution (10 mg/ml in chloroform)
under a stream of nitrogen. TBS solution is added to the dried lipid film

to give a final concentration of 0.5 mg lipid/ml suspension. The mixture is sonicated for 15 min at 40°C with a tip probe sonicator at an amplitude of 16 microns (MSE ultrasonic desintegrator MK 2). Alternatively, sonication can be carried out in a thermostated bath-type sonicator at 40°C for at least 30 min. The sonicator bath is filled with a solution of 1% sodium dodecyl sulfate in water. It should be noted that the velocity of DMPC-induced vesicle formation is strongly dependent on the amount of added lipid. Therefore, to obtain reproducible release rates, the lipid concentration should be determined and, if necessary, adjusted.

C. Extraction of Lipids

The lipid concentration can be determined as follows: 0.5 ml (= 1 volume) of the lipid suspension are pipetted into a 10 ml reagent tube, 2.2 ml (= 4.3 vol) of chloroform-methanol 5:8 (v/v) are added and the mixture homogenized with a vortex mixer. The mixture should still show one phase; if not, a few drops of methanol are added. After 15 min, 1.9 ml (= 3.8 vol) of chloroform and 0.5 ml (= 1 vol) of water are added, followed by vigorous mixing. Separation of the two phases is then obtained by centrifugation for 15 min at about 2500 rpm in the clinical centrifuge. The aqueous upper phase and the interfacial film are completely but carefully removed by aspiration. An aliquot of the lower (chloroform) phase is transferred into an acid-cleaned 10 ml test tube and dried under a stream of nitrogen.

D. Determination of Lipid Phosphorus

An amount of 0.65 ml of perchloric acid is added to the dry film of extracted lipid, the tube is closed with a (clean) glass marble and heated to 180° – 210°C for at least 20 min. After cooling, 3.3 ml water, 0.5 ml of a 2.5% solution of ammoniumheptamolybdate in water and 0.5 ml of a 10% solution of ascorbic acid in water are added (the solutions have to be stored in the refrigerator). The tube is again sealed with a marble and heated for 5 min in a boiling water bath. After cooling, the absorbance at 797 nm is read against a blank and the amount of phospholipid calculated, using KH_2PO_4 as a standard for the calibration curve.

E. Incubation of Erythrocytes with DMPC

Two ml of packed erythrocytes are suspended in 18 ml of lipid dispersion and incubated with gentle stirring at 30°C. After appropriate incubation

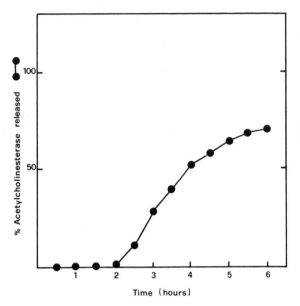

Fig. 1. Release of membrane vesicles. Erythrocytes were incubated as described in the text. Acetylcholinesterase activity in the supernatant of the low speed centrifugation step was taken as a measure for vesicle release

times the mixture (or an aliquot) is centrifuged at 3000 rpm for 20 min. The released vesicles, which contain acetylcholinesterase, can be collected in the supernatant. The velocity of vesicle release is conveniently monitored by determination of acetylcholinesterase activity in the supernatant fraction after various incubation times (Fig. 1): 3 ml of assay mixture (see Chemicals) are filled into a cuvette and the reaction is started by addition of vesicle suspension (e.g., 50 μl). The change in absorbance is monitored at 412 nm and registered with a recorder. The activity is calculated using a molar extinction coefficient of 13,600 for the reduced DTNB as follows:

$$\frac{IU}{ml \text{ specimen solution}} = \frac{\Delta A\,412}{min} \; \frac{FV}{SV} \; \frac{7.35}{100}$$

FV: Final volume (ml) of assay; PV: volume (ml) of specimen

An increase of the incubation temperature results in a faster vesicle formation (Fig. 2). Incubations should not be carried out for more than 6 h at 30°C, because prolonged incubation causes hemolysis.

Vesicles obtained in the supernatant of the low speed centrifugation can be concentrated and separated from the bulk of DMPC by centrifugation at 15,000 rpm (30,000 g) and 30°C for 30 min. The pellet is washed twice by resuspension in 20 ml of isotonic TBS. The pellet obtained after the last centrifugation step is resuspended in 2 ml of TBS.

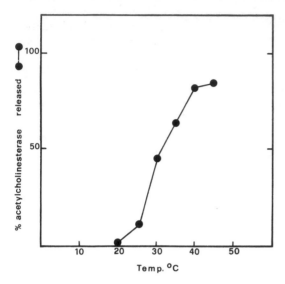

Fig. 2. Temperature-dependence of vesicle release. Incubation of erythrocytes was carried out at different temperatures for 4 h. Acetylcholinesterase activity in the supernatant of the low speed centrifugation step was taken as a measure for vesicle release

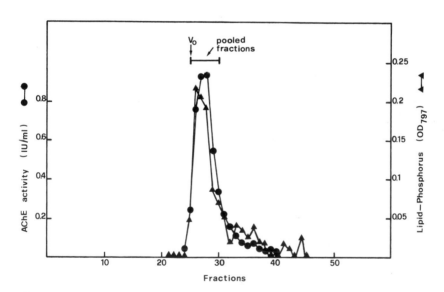

Fig. 3. Column chromatography of concentrated vesicle preparation. Vesicles were prepared, concentrated and layered onto the Sepharose 4 B column as described in the text. The fractions were assayed for acetylcholinesterase activity and lipid phosphorus. The pooled fractions indicated were used for analysis of phospholipid composition. Vo indicates the void volume of the column

Alternatively, purification of the vesicles is performed by gel filtration on Sepharose 4B. For this purpose, the vesicles obtained by low speed centrifugation are concentrated by centrifugation at 15,000 rpm for 30 min. The pellet is resuspended in 2 ml TBS and applied onto a Sepharose 4B column (2.5 × 10 cm) thermostated at 30°C. The vesicles are eluted in the void volume of the column (Fig. 3). The preparation procedures are summarized in Fig. 4.

The pooled peak fractions as well as the vesicles purified by repeated centrifugation are essentially devoid of DMPC liposomes, which is documented by freeze-fracture electron microscopy (Fig. 5) and a determination of the phospholipid composition (Table 1). Phospholipids can be separated by two-dimensional thin-layer chromatography, as described in Chapter 1.3. The quantitative distribution of the phospholipids can be obtained by scraping the individual spots from the plate (with a razor blade) into test tubes and the subsequent determination of lipid phosphorus in each tube. The method for phosphorus determination has been described earlier in this chapter (Sect. III.D).

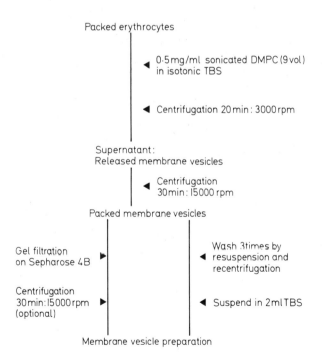

Fig. 4. Preparation of membrane vesicles from human erythrocytes

Fig. 5. Freeze-fracture electron microscopy of purified vesicle preparation. *The bar* indicates 100 nm

Table 1. Comparison of phospholipid composition in released vesicles, erythrocytes after incubation with dimyristoylphosphatidylcholine and intact erythrocytes. Lipids were separated and determined as described in the text. Data are expressed as mol% of total phospholipid fraction (mean of three experiments ± S.D.)

Component	Vesicles	Incubated erythrocytes	Erythrocytes [a]
Sphingomyelin	27.9 ± 4.1	23.3 ± 4.4	25.8 ± 1.7
Phosphatidylcholine	42.4 ± 2.6	43.6 ± 5.8	28.3 ± 2.1
Phosphatidylethanolamine	21.3 ± 1.3	24.2 ± 3.9	26.7 ± 1.0
Phosphatidylserine	8.2 ± 1.9	8.9 ± 1.5	12.7 ± 1.3

a   From van Deenen and de Gier (1974)

The erythrocyte membrane vesicle preparations obtained contain the major erythrocyte phospholipid classes in a similar quantitative relation as found in the native membrane. Band 3 protein, glycophorin and acetylcholinesterase are present, while spectrin is not found. Hemoglobin is included in the lumen of the vesicles. Incubation does not result in ATP-depletion of the erythrocytes (Ott et al. 1981).

References

Lutz HU, Lio S-C, Palek J (1977) J Cell Biol 73:548—560
Ott P, Hope MJ, Verkleij AJ, Roelofsen B, Brodbeck U, van Deenen LLM (1981) Biochim Biophys Acta 641:79—87
Schwoch G, Passow H (1973) Mol Cell Biochem 2: 197—217
Steck TL (1974) Methods Membrane Biol 2: 245—281
van Deenen LLM, de Gier J (1974) In: Surgenor D (ed) The red blood cell, 2nd ed, vol 1. Academic Press, New York, pp 147—211

Purification by Affinity Chromatography of Red Cell Membrane Acetylcholinesterase

U. BRODBECK, R. GENTINETTA, and P. OTT

I. Introduction

Over the past the methods for purifying integral membrane proteins have advanced considerably. In particular the onset of affinity chromatography (Jakoby and Wilchek 1974) has brought new incentives to this particular class of proteins. In order to be deployed successfully the following criteria have to be met: (A) A high affinity ligand covalently linked to an insoluble support via a spacer arm must warrant biospecific adsorption. After removal of all undesired protein, desorption should be carried out using similar biospecificity. (B) The amphiphilic membrane protein must be kept in solution throughout the purification procedure by the use of a suitable detergent. Alternatively membrane proteins may be solubilized by proteolytic enzymes, chaotropic ions or extraction with organic solvents.

The present experiment illustrates the usefulness and power of the method. It shows that a minor compound accounting for less than 0.1% of all membrane proteins can be successfully purified essentially in a one-step purification scheme (Berman and Young 1971; Ott et al. 1975). It relies on the fact that aromatic quaternary nitrogen-containing compounds are powerful reversible inhibitors of acetylcholinesterase and that a sufficiently long spacer arm warrants tight binding of the high affinity ligand to the micellarized enzyme (Fig. 1). On the other hand, the positively charged group covalently bound to an insoluble support, renders any resin a strong anion exchanger which primarily will unspecifically bind any negatively charged protein. This apparent disadvantage is, however, overcome at sufficiently high ionic strength (i.e., in 1 M salt solutions, where unspecific binding is eliminated). Given enough time to shift the equilibrium toward the much tighter binding of acetylcholinesterase, all contaminating proteins can then be eluted from the resin at this ionic strength (Fig. 2). Under these conditions, only a batch-type operation warrants successful adsorption of the enzyme, as ordinary column chromatography would only adsorb the majority of contaminating protein. Using this method, solubilized acetylcholinesterase can be purified either from isolated red cell membrane (ghosts) or from a total red cell hemolysate. Owing to large differences in the initial specific activity of

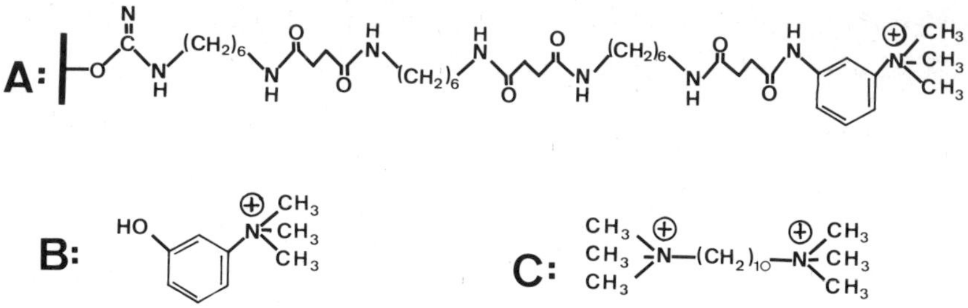

Fig. 1. Affinity resin *(A)* for the purification of human red cell membrane acetylcholinesterase and displacing high affinity ligands, edrophonium (= tensilon, *B*) or decamethonium *(C)*

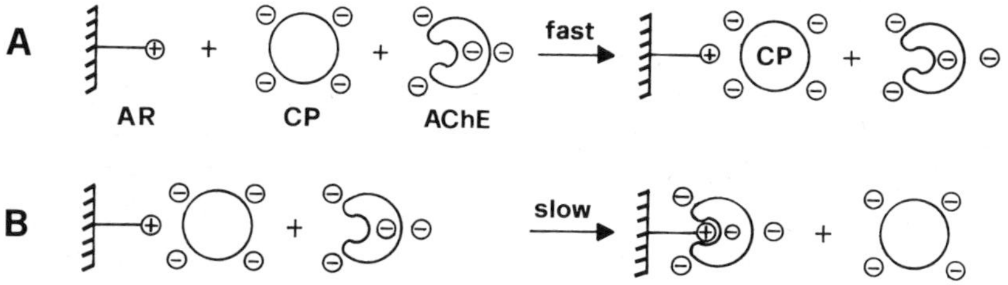

Fig. 2. Binding of acetylcholinesterase *(AChE)* to high affinity resin *(AR)* in presence of a vast excess of contaminating protein *(CP)*. In a fast reaction *(A)* excess of contaminating protein will bind unspecifically to the cationic resin, thus rendering the binding of acetylcholinesterase essentially to zero. Then a slow shift in equilibrium is seen *(B)*, after which the enzyme becomes tightly bound to the resin. Thereafter contaminating protein can be eluted at high ionic strength

Table 1. Comparison of conditions in the purification of acetylcholinesterase depending on the starting material

Starting material	Specific enzyme activity	Time aquired for enzyme to bind to high affinity resin	Comments
	IU/mg protein	h	
Isolated membrane (ghosts)	>2	12–18	Isolation of large amounts of ghosts is time consuming; affords large capacity centrifuge
Red cell hemolysate	<0.02	48–72	Suitable to produce mg-amounts of enzyme, affords larger amounts of high affinity resin

the enzyme, differences in times are encountered during which the change in the above-mentioned equilibrium takes place (Table 1).

In the present experiment the isolation of enzyme using red cell hemolysate as starting material is described which, comparing effort and time, will yield larger amounts of pure enzyme than a ghost preparation. The procedure outlined is, however, equally suited for the latter case. The enzyme is solubilized by the nonionic detergent Triton X-100 which has to be present throughout the purification procedure in amounts exceeding the critical micellar concentration. It could be shown that the pure enzyme in presence of detergents above their critical micellar concentration is a dimer with two subunits interlinked by at least one disulfide bridge (Römer-Lüthi et al. 1979). This enzyme form sediments upon density gradient centrifugation as a homogenous 6.5 S entity. Its catalytic activity is absolutely dependent on the presence of a stabilizing amphiphile such as detergent, a lyso- or a phospholipid molecule (Wiedmer et al. 1979).

If desired the pure enzyme may be freed of the solubilizing detergent (e.g., Triton X-100). This then leads to self micellarization of the enzyme at a protein concentration above 2.5 μg per ml which is a typical phenomenon observed with amphiphilic membrane proteins. Sucrose density gradient centrifugation offers an easy mean to demonstrate this property (Ott and Brodbeck 1978; Wiedmer et al. 1979).

The enzyme is readily introduced into lipid vesicles (Hall and Brodbeck 1978; Frenkel et al. 1980; Römer-Lüthi et al. 1980). Arrhenius plots established with enzyme incorporated into lipid bilayers of different fluidity clearly showed that the enzyme activity is modulated by the phospholipid molecules surrounding the protein (Frenkel et al. 1980).

II. Equipment, Chemicals, Solutions, and Biological Material

A. Equipment

assembled equipment as shown in Fig. 3
Buchner funnel
vacuum desiccator
250 ml round bottomed flask
100 ml round bottomed flask
double-necked round-bottomed flask equipped with reflux condenser and
    dropping funnel
drying tube
large capacity centrifuge (e.g., MSE cool spin with 6 X 1 L swing-out rotor)

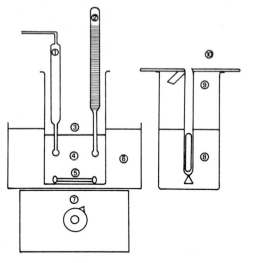

Fig. 3. Equipment used in the synthesis of the high affinity resin. *1* Electrode connected to a pH-meter; *2* thermometer; *3* 500 ml beaker; *4* suspension of Sepharose 4B; *5* dumbbell-type magnetic stirring bar; *6* ice bath; *7* magnetic stirring motor; *8* alternative stirring device using ordinary stirring bar entrapped in a narrow dialysis casing *(9)* held in position by a glass rod *(10)* or a metal clamp

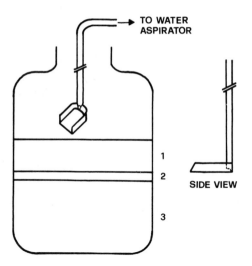

Fig. 4. Spoon-type aspirator to remove supernatant *(1)* and buffy coat layer *(2)* on top of red cell sediment *(3)*

ordinary laboratory centrifuge (e.g., MSE high spin 21 with 6 × 500 ml angle rotor)

high speed centrifuge (e.g., MSE Superspeed 65 with a 10 × 10 ml angle rotor)

spoon type aspirator device as shown in Fig. 4

overhead paddle type stirring device

recording spectrophotometer

equipment for column chromatography, fraction collector and pump

ultrafiltration cell, 50 ml (Amicon equipped with PM 10 membrane and a concentration/dialysis selector CDS 10 connected to a 800 ml reservoir)

B. Chemicals

sepharose 4-B, 100 ml bed volume (Pharmacia)
cyanogen bromide (Fluka Nr. 16771, 24 g)
1,6-diaminohexane (Fluka Nr. 33000, 72 g)
succinic anhydride (Fluka Nr. 14090, 60 g)
water-soluble carbodiimide: N-cyclohexyl-N′-[2-morpholinyl)-ethyl]-carbo-
    diimide-methyl-p-toluenesulfonate (Fluka Nr. 29469, 30 g)
N,N-dimethyl-m-phenylene diamine dihydrochloride (Eastman Nr. 9145,
    11 g)
2,4,6-trinitrobenzenesulfonic acid (Eastman Nr. 8746, 5 g)
acetic anhydride (50 g)
acetone
ethanol
methyl iodide (14 g)
Trito X-100 (Röhm and Haas)
hydroxylapatite (e.g., Biogel HTP, BioRad Laboratories)
edrophonium chloride (Roche)
decamethonium bromide (Fluka Nr. 30520)

C. Solutions

1. For Chemical Synthesis

NaOH 1, 3 and 10 N
HCl 1 N and fuming
H_2SO_4 concentrated
$NaHCO_3$ 0.1 N
Na_2CO_3 0.1 N
$Na_2B_4O_7$ saturated solution (50 ml)
2,4,6-trinitrobenzenesulfonic acid (3% in saturated sodium tetraborate solu-
    tion
6 M urea in H_2O

2. For Enzyme Purification (Letters a-h as Indicated in Text)

a) NaCl, 1% in H_2O containing 0.05% NaN_3 (6 L)
b) Triton X-100, 3% in H_2O containing 45 mM EDTA and 0.075% NaN_3 (3 L)
c) 20 mM Tris-HCl buffer, pH 7.4, containing 0.1 M NaCl, 0.05% NaN_3,
    30 mM EDTA and 1% Triton X-100 (2 L)

d) 20 mM Tris-HCl buffer, pH 7.4, containing 1.0 M NaCl, 0.05% NaN_3, 30 mM EDTA and 1% Triton X-100 (2 L)
e) 20 mM Tris-HCl buffer, pH 7.4, containing 0.8 M NaCl, 0.05% NaN_3, 30 mM EDTA and 0.05% Triton X-100 (1 L)
f) same as buffer (e) containing in addition 20 mM Edrophonium chloride (Roche, 200 ml). If not available, decamethonium bromide (20 mM) may be used instead
g) 10 mM sodium phosphate buffer, pH 6.0, containing 0.05% NaN_3 (500 ml). Correct pH of phosphate buffer at 4°C is of importance!
h) 200 mM sodium phosphate buffer, pH 7.4, containing 0.05% NaN_3 (200 ml). Kept at room temperature, buffer will crystallize in the cold

3. For Enzyme Assay

Solution containing 1 mM acetylthiocholine, 0.125 mM 5,5'-dithio-bis (2-nitrobenzoic acid (= Ellman's reagent) and 0.05% Triton X-100 in 100 mM sodium phosphate buffer, pH 7.4.

D. Biological Material

10 red blood cell sediments, fresh or outdated from 400 ml whole blood each as obtained through a local blood bank.

III. Experimental Procedure

A. Synthesis of a High Affinity Ligand: N,N,N-trimethylammonium-m-phenylenediamine

Starting from the commercially available N,N-dimethyl-m-phenylenediamine the synthesis is carried out as depicted in Fig. 5.

Reaction batch: 10.5 g (50 m Mol) N,N-dimethyl-m-phenylenediamine
                  50   g (500 m Mol) acetic anhydride
                  0.5 ml concentrated H_2SO_4

Place the diamine in a 250 ml round-bottomed flask stoppered with a drying tube, add the sulfuric acid and one half of the acetic anhydride. Cautiously warm to 70°C and slowly add the remaining acetic anhydride. Let the

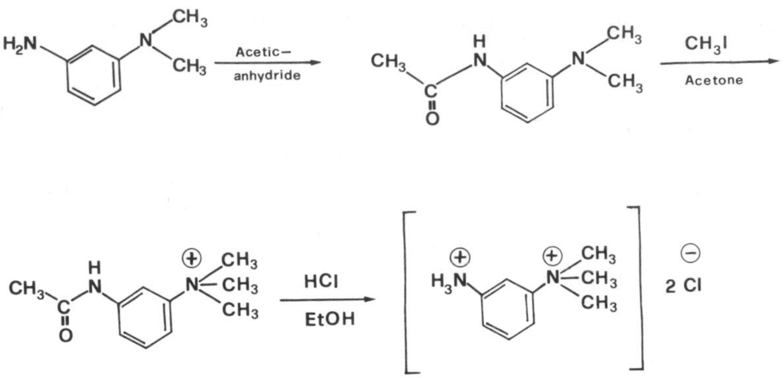

Fig. 5. Synthesis of high affinity ligand starting from a commercially available chemical

stoppered solution stand until its temperature becomes ambient. Then cool in ice and slowly pour the solution into an ice-cooled 250 ml Erlenmeyer flask containing 10 ml of water. Keep the temperature as low as possible. Slowly add 10 N NaOH to a pH of 12.0 (approximately 60 ml) and keep the temperature always below 35°C. In this step the yellow-colored N-acetyl-m-N,N-dimethylphenylenediamine will precipitate. Keep the mixture at 4°C overnight. Filter on a Buchner funnel and wash the precipitate with cold 0.1 N NaHCO₃ in order to remove the coprecipitated sodium acetate.

Place the precipitate in a 250 ml two-necked round-bottomed flask equipped with a reflux condenser. Add 100 ml of acetone and reflux until the precipitate has dissolved completely. Place a dropping funnel containing 14 g ($\approx$ 100 m Mol) methyl iodide into the side arm and slowly add the methyl iodide to the solution. Shake occasionally or stirr with a stirring device while the quaternary amine will precipitate. Keep the mixture at 4°C overnight and filter the product through a Buchner funnel. Place the precipitate in a 100 ml single-neck round-bottomed flask equipped with a reflux condenser and dissolve it in a minimum of a solution of ethanol and concentrated HCl (1:1 v/v). Reflux the yellow solution for 2 h and then slowly add acetone until the product precipitates. Let the flask stand until the solution assumes room temperature and then keep at 4°C overnight. Filter the precipitate on a Buchner funnel and wash with cold acetone. Recrystallize by dissolving the product again in hot ethanol/HCl (1:1 v/v) as above and add acetone until a faint precipitate appears. Let stand overnight at 4°C and filter the product. MP (uncorrected) 196°–198°C.

B. Synthesis of High Affinity Resin

The synthesis is carried out as outlined in Fig. 5. Assemble equipment as shown in Fig. 3 in a well-ventilated hood. Place 100 ml of bed volume Sepharose 4-B (normal or cross-linked) in the beaker and add 100 ml of H_2O. Gently stir the suspension using a dumbbell-type magnetic stirring bar or alternatively an ordinary stirring bar entrapped in a dialyzing casing and mounted as shown in Fig. 3. Do not use an ordinary stirring bar spinning on the bottim as it may grind the resin to fine particles, thus causing losses and reduced flow rates.

1. Activation by Canogen Bromide

Slowly add 24 g cyanogen bromide[1] in small portions and keep the pH between 10.0 and 10.3 by dropwise adding 10 N NaOH.
The temperature must not exceed 30°C. Keep adding NaOH until the pH remains constant (Usually after 3–4 h). Filter the resin through a Buchner funnel and rinse immediately with large amounts (600 ml) of cold 0.1 N sodium carbonate buffer, pH 10. Disconnect the suction and stir in immediately a solution of 23.5 g 1,6-diaminohexane in 50 ml of H_2O previously adjusted to pH 10 with concentrated HCl (affords cooling). Rinse suspension into a beaker and stir overnight, using an overhead stirring device.
Wash the resin thoroughly with water on a Buchner funnel and resuspend it in 100 ml H_2O. Carry out the color test for successful amino-coupling as follows. To a small amount of Sepharose suspension (0.2–0.5 ml) add 1 ml of a saturated solution of sodium tetraborate and 3 drops of a solution of 2,4,6-trinitrobenzene sulfonic acid in saturated sodium tetraborate. Shake the solution and let the Sepharose settle out. Owing to the presence of a primary amino group the beads will appear orange-colored which is best seen by comparison with underived Sepharose.

2. Succinylation

Degas the derived Sepharose in a vacuum desiccator and again place it in the same experimental set up. Keep the slurry at 4°C and adjust the pH to 8.0. Add 20 g succinic anhydride in small portions and keep the pH between 7.8 and 8.2 by dropwise addition of 3 N NaOH. Continue until the pH stays

---

1 Weigh out cyanogen bromide in a well-ventilated hood as this chemical is a lacrimator. Avoid contact with the skin as cyanogen bromide is highly aggressive. Once at alkaline pH, it may be handled without special precaution

constant (3–4 h). Keep stirring at 4°C for 20 h. Wash the resin well with water and carry out the color test as indicated above. If succinylation has been carried out successfully the beads will now appear colorless.

3. Chain Elongation

Again degas the Sepharose and using the same equipment adjust the pH of the slurry to 4.75 with 1 N HCl. Keep the temperature between 10° and 15°C and add 23.5 g 1,6-diaminohexane in water previously adjusted to pH 4.75 using concentrated HCl (affords cooling). Add in small portions 10 g of the water-soluble carbodiimide and keep the pH between 4.7 and 4.8 using 1 N HCl. When the reaction is completed, continue stirring in the cold for another 20 h. Carry out the color test; the beads will now appear orange.

Continue the chain elongation by alternatively coupling succinic anhydride and 1,6-diaminohexane until the chain equals to the structure shown in Fig. 1A.

4. Coupling the High Affinity Ligand

Degas the Sepharose and add to it 2.5 g of the high affinity ligand dissolved in a few ml of water. Adjust the pH to 5.0, slowly add 10 g of the water soluble carbodiimide and keep the pH between 4.9 and 5.1 until the reaction is completed (usually within 3 h). Continue stirring for 20 h in the cold. Wash the affinity resin with plenty of water, then place it into a large enough column and percolate it with 3 volumes of 6 M urea followed by solution (c) if the resin is used immediately for enzyme preparation. Alternatively if the resin is to be kept, rinse well with water containing 0.05% NaN$_3$ and store at 4°C.

C. Enzyme Purification

1. Washing of Erythrocyte and Hemolysis

The purification may conveniently be started on a Thursday using 10 or more red blood cell sediments (fresh or outdated) as obtained through a local blood bank. All operations are carried out in the cold at 4°C and all solutions should be precooled to 4°C [exception: solution (h)].

The pooled red cell sediments (approximately 2–3 L) are spun at 3000 rpm for 20 min in a large capacity centrifuge. The supernatant plasma and buffy coat are generously removed using a spoon-type aspirator (Fig. 4). To the remaining sediment (1.5–2 L) an equal volume of solution (a) is added and gently mixed. The cells are sedimented as above and the buffy coat layer removed. Repeat this washing process twice more in order to ensure an almost complete removal of leukocytes and platelets (leukocytes contain many proteases which could endanger the subsequent enzyme purification).

The washed erythrocytes are pooled and the volume (usually between 1 and 1.5 L) is noted. Two volumes of solution (b) are added to give a final Triton X-100 concentration of 2%. The mixture is then stirred for at least 3 h (or overnight, if convenient) preferably using a slowly rotating paddle-type stirring device (beware of foam formation). Remove a 5–10 ml aliquot (X) and store it in the cold.

2. Affinity Chromatography

Degas at least 50 ml bed volume of the affinity resin in a desiccator using an ordinary water aspirator until no more air bubbles appear. If in stock, double the volume of affinity resin which ensures faster binding and higher yields. Pour the resin into the hemolysate and slowly continue stirring for 48 to 72 h. During this time the use of a paddle-type stirring device is imperative as a magnetic stirring bar will grind and disintegrate the affinity resin. Thereafter remove another 5–10 ml aliquot (Y) and centrifuge both (X) and (Y) for 1 h at 100,000 g.

Measure acetylcholinesterase activity in the supernatant fraction and calculate the amount of enzyme bound to the affinity resin (X − Y). Usually between 60%–70% of enzyme become attached within 72 h. As the marginal increase observed thereafter does not warrant a much longer incubation, stirring is discontinued and the affinity resin allowed to settle. Filter the solution through a sheet of nylon, fine enough to retain the resin. Distribute the resin among 2500 ml centrifuge beakers and add solution (c) to near the top of the beakers. Gently mix and centrifuge at 3500 rpm for 20 min (low speed centrifugation is essential to avoid dense packing of the affinity resin that would cause a gummy residue). Carefully decant the supernatant and retain all resin in the centrifuge beaker. Combine the supernatants and check for enzyme activity (should amount to less than 1% of that adsorbed on the resin). Repeat washing of the resin until the supernatant appears almost free of hemoglobin. Then continue washing with solution (d) until the supernanant appears faintly red-colored.

Thereafter combine the affinity resin, degas it and place it into a column (2.5 to 4 cm diameter, depending on the amount used) and percolate it with solution (d) at a rate of 20–30 ml/h using hydrostatic pressure only, until no more protein is detectable in the eluate (may take 4–5 days). The hydrostatic pressure should not exceed 60 cm H_2O. Reduce the Triton X-100 and salt concentration by percolating solution (e) for 12 h at a pumped rate of 12 ml/h and then displace the enzyme from the affinity resin by applying solution (f). If the enzyme is to be kept in this detergent-containing solution, then the displacing ligand has to be removed by extensive dialysis against solution (e) devoid of EDTA prior to a definite assessment of activity and yield.

3. Removal of Triton X-100

If, on the other hand, the enzyme is to be freed from excess Triton X-100, proceed as follows without prior dialysis. The enzyme in solution (f) is placed in a 50 ml Amicon ultrafiltration cell fitted with a PM 10 membrane. Applying about 1 atm N_2-pressure the solution is first concentrated to about 20 ml and then diafiltrated with solution (g) (approximately 200 ml) until conductivity and pH in the eluate match the values for solution (g). In this step dialysis is not recommended as a prolonged exposure of the enzyme to pH 6 causes loss of activity.

The enzyme is applied onto a column (approximately 1 X 15 cm) of hydroxylapatite (e.g., Biogel HTP) previously equilibrated with solution (g). Percolate solution (g) at a pumped rate of 12 ml/h through the column until A_{280} is reduced to zero. Check for activity in the eluate in order to assure that the enzyme remains bound to the hydroxylapatite throughout the washing. Then apply solution (h) which quantitatively displaces the enzyme from the column. [Solution (h) must be kept at ambient temperature in order to avoid crystallization of the buffer.]

Enzyme Assay. Acetylchonlinesterase activity is measured according to the method of Ellman et al. (1961). The reaction is followd spectrophotometrically in a 3 ml cuvette by the increase in absorbance at 412 nm. Enzyme activity (IU/ml) is calculated according to the general formula given below, using a molar absorption coefficient of 13,600 for the nitrobenzoate anion produced in the reaction.

$$\frac{IU}{ml \text{ of enzyme solution}} = \frac{\Delta A_{412}}{minute} \cdot \frac{FV}{SV} \cdot \frac{7.35}{100}$$

in which FV equals the final volume (ml) and SV the volume (ml) of specimen (enzyme) in the assay.

IV. Recommended Time Schedule for Enzyme Purification

Thursday: Obtain and wash the red cell sediments. Hemolyse overnight.
Friday: Add affinity resin to the heolysate and stir until Monday morning.
Monday: Wash the affinity resin and continue throughout the week as outlined in the text.

V. Comments

This purification procedure yields an enzyme with an average specific activity of 4000 IU/mg protein. It can be increased to about 6000 by a second affinity chromatography step using a column of 20 ml bed volume resin. Herein the column is washed and the enzyme eluted as indicated above. The enzyme obtained after 1 chromatographic step is suitable for reconstitution and crosslinking experiments outlined on p. 163. If the amphiphile dependency (Wiedmer et al. 1979) is to be studied, the Triton X-100 depleted, aggregated enzyme is dissociated into the dimeric form using 1.5% β-D-octylglucoside as detergent. The 6.5 S enzyme is then isolated by sucrose density gradient centrifugation in a buffer containing 1.5% β-D-octylglucoside.

Acknowledgment. Supported by grants no. 3.032-0.76 and 3.577.79 of the Swiss National Science Foundation.

References

Berman JD, Young M (1971) Proc Natl Acad Sci USA 68:395–398
Ellmann GL, Courtney DK, Andres V, Featherstone RM (1961) Biochem Pharmacol 7: 88–95
Frenkel EJ, Roelofsen B, Brodbeck U, van Deenen LLM, Ott P (1980) Eur J Biochem 109:377–382
Hall ER, Brodbeck U (1978) Eur J Biochem 89:159–167
Jakoby WB, Wilchek M (eds) (1974) Methods in enzymology, vol 34. Academic Press, New York
Ott P, Brodbeck U (1978) Eur J Biochem 88:119–125
Ott P, Jenny B, Brodbeck U (1975) Eur J Biochem 57:469–480
Römer-Lüthi CR, Hajdu J, Brodbeck U (1979) Hoppe-Seyler's Z Physiol Chem 360: 929–934
Römer-Lüthi CR, Ott P, Brodbeck U (1980) Biochim Biophys Acta 601:123–133
Wiedmer T, Di Francesco C, Brodbeck U (1979) Eur J Biochem 102:59–64

Purification of Cytochrome c Oxidase by Affinity Chromatography

A. AZZI, K. BILL, C. BROGER, R.P. CASEY, and V. RIESS

I. Introduction and Aims

Cytochrome c oxidase (EC 1.9.3.1) can be purified from a large number of species and tissues with techniques based on the detergent solubilization of this membrane-bound enzyme, and its subsequent salt precipitation (Azzi 1981; Azzi and Casey 1979).

In general one to two days are necessary with the traditional procedures in order to purify the enzyme, while with the technique described here cytochrome c oxidase can be obtained in seven hours and with an almost automated procedure (Bill et al. 1980).

The principle of this technique is that of covalently binding cytochrome c to a resin, leaving however the ϵ-amino groups centered on lysin 13 free for reacting with cytochrome c oxidase (Rieder and Bosshard 1978). This cannot be achieved using CNBr-activated Sepharose 4B since proteins are bound to the CNBr activated Sepharose via ϵ-amino groups.

Yeast cytochrome c, which at position 103 contains a cystein residue, can be bound to a Thiol-activated Speharose 4B by forming a disulfide bond between the cytochrome and the resin. In this case the lysine ϵ-amino groups can still interact with cytochrome c oxidase and promote its binding. The complex cytochrome c-cytochrome c oxidase realized through ionic interactions can be dissociated by salts (Jacobs and Sanadi 1960).

II. Equipment and Solutions

A. Equipment

refrigerated centrifuge (Sorvall RC-5)
recording spectrophotometer (Aminco DW-2)
fraction collector (LKB)
pipettes
sintered glass filter (G3)
two columns for chromatography (10 X 1 cm)
peristaltic pump (LKB)

B. Solutions

beef-heart mitochondria (frozen)
thiol-activated Sepharose 4B (4 g in 30 ml water) (Pharmacia)
Sephadex G25 (2 g in 10 ml water) (Pharmacia)
medium 0.05 M Tris-HCl, 1 mM EDTA, pH 7.4
cytochrome c (horse heart) 50 mg/ml (Sigma)
cytochrome c (yeast) 50 mg/ml (Sigma)
Tween 80 20% in H_2O
Triton X-100 20% in II_2O (w/v) (Fluka)
NaCl, 1 M
dithionite, solid
cysteine 250 mM

The type of instruments and the origin of chemicals used in this experiment is only indicative, and can be substituted with any other equivalent ones.

III. Experimental Procedures

A. Preparation of the Affinity Column

Swell the thiol Sepharose 4B in 30 ml distilled water for 30 min and wash on a sintered glass filter (G3) with 1.2 l distilled water. 12 ml hydrated gel are obtained, which will be divided into two equal portions and packed into the two columns, which will be called A and B.

Column A is loaded with the solution of yeast cytochrome c and its two ends connected through a peristaltic pump and the cytochrome c solution circulated for 8 h at 4°C. In this time more than 90% of the added cytochrome c is bound.

Column B is washed with 100 ml of 0.25 M cysteine solution. Both columns are washed with 50 ml buffer and equilibrated with 50 ml of the same buffer containing 0.1% Triton X-100.

B. Purification of Cytochrome c Oxidase

Mitochondria are diluted with the buffer to 2 mg/ml and 1% Triton X-100 is added. After centrifugation at 28,000 g for 40 min a supernatant is obtained. Both columns are loaded each with 30 ml of supernatant. Elution is obtained with the buffer containing 0.1% Triton X-100 at a pump flow rate of 12 ml/h. 3 ml fractions are collected. A calculation of the content of cytochromes in the eluted samples indicates that nothing is retained by the deactivated (cystein treated) thiol Sepharose 4B column (column B), and thus this column will not be further used.

After 15 samples are collected, the elution of column A is continued with the same medium brought to 50 mM in NaCl (using the 1 M solution). Cytochrome oxidase is eluted in the next 30 ml. After 100 ml of total eluate is collected, NaCl is added to the elution buffer to bring its concentration to 100 mM.

At this point the b-c_1 complex is eluted.

C. Measurements

Spectra. Take a spectrum (reduced with dithionite *minus* oxidized) of representative fractions of the eluate and of the originally Triton X-100 solubilized material in the spectrophotometer from 400 to 650 nm, using undiluted samples.

Enzyme Activtiy. Reduced cytochrome c is obtained by adding dithionite (a few grains) to the horse heart cytochrome c solution. The reduced sample is applied to a Pasteur pipette filled with the Sephadex-G25 resin and eluted with the medium to remove the excess of reducing agent and its products. 10 μM cytochrome c (reduced) is added to 2 ml of medium containing 0.5% Tween 80. The spectrophotometer is set to 550–540 nm, aliquots of cytochrome c oxidase-containing samples added and the rate of decrease of absorbance measured.

IV. Calculations

Enzyme Purity. Measure the difference in absorbances at the following wavelength pair and calculate, using the Lambert-Beer equation:

$$\Delta A = c \times d \times \epsilon$$

and the given extinction coefficient, ϵ, the concentration of the different cytochromes

Cytochrome	λ	ϵ_{mM}	ΔA	Concentration
a–a_3	605–630 nm	16.0		
b	562–575 nm	20.0		
c_1	554–540 nm	17.1		
Protein	Lowry et al. (1951)	–	–	

The purity of the enzyme can be expressed in μmol of heme $a-a_3$ per g protein and has a maximum value of 14—16.

Enzymatic Activity. It is calculated from the following expression:

$$M.A. = \Delta A/(0.019.s. \mu M \text{ heme a})$$

where 0.019 represents the μM extinction of cytochrome c at 550—540 nm.

The molecular activity (M.A.) of cytochrome c oxidase can be expected to be in the order of 100—200 μmol cytochrome c oxidized per second per μmol heme a.

References

Azzi A. (1981) Biochim Biophys Acta, in press
Azzi A, Casey RP (1979) Mol Cell Biochem 28:169—184
Bill K, Casey RP, Broger C, Azzi A (1980) FEBS Lett 120:248—250
Jacobs EE, Sanadi DR (1960) J Biol Chem 235:531—534
Lowry OH, Rosebrough NJ, Farr AL, Randall RJ (1951) J Biol Chem 193:265—275
Rieder R, Bosshard R (1978) FEBS Lett 92:223—226

Reconstitution

Reconstitution of Pure Acetylcholine Receptor

H. LÜDI and U. BRODBECK

I. Introduction and Aims

The procedure for the isolation and the purification of the acetylcholine receptor from the electric organ of *Torpedo marmorata* has been described in Chapter 2.3. Reconstitution of the acetylcholine receptor in artificial lipid bilayers has been reported by several authors (Changeux et al. 1979; Howell et al. 1978; Huganier et al. 1979; Schiebler and Hucho 1978; Wu and Raftery 1979).

In this chapter, a method is described which allows the reconstitution of the acetylcholine receptor into unilamellar lipid bilayers directly from Triton X-100-containing solutions. This method avoids undesirable effects of ionic detergents as well as of sonication on the purified receptor protein (Elliot and Raftery 1979; Eldefrawi et al. 1978; Heidmann and Changeux 1978). The only compounds for the reconstitution are: phosphatidylcholine as vesicle-forming lipid, Triton X-100 as nonionic detergent to resolve the phospholipids and to desaggregate the receptor protein, and Amberlite XAD-2 as detergent removing agent.

II. Equipment, Solutions and Materials

A. Equipment

equipment for the determination of the number of α-bungarotoxin binding
  sites: see p. 70
MSE Prespin 65 centrifuge, 6 × 14 ml swing-out rotor
Sepharose 4B column 2.5 × 30 cm
glass counting vial for stirring reconstitution medium

B. Solutions and Materials

Methanol p.a. (Merck)
Triton X-100 (Röhm and Haas)

[^{3}H]-Triton X-100, specific activity 0.246 mCi/g (Röhm and Haas)
L-α-Lecithin ex egg (Koch-Light Laboratories Ltd., England)
pure acetylcholine receptor: specific activity 10.5 nmol α-bungarotoxin-binding sites/mg protein (for purification see previous chapter)
[^{125}I]-labeled α-bungarotoxin (for purification and iodination see previous chapter)
10 mM Tris-HCl, pH 7.4 buffer containing 0.2 M KCl and 0.5% Triton X-100 (w/v)
Amberlite XAD-2 (300–1000 μm) (Serva)
Sepharose 4B (Pharmacia)

III. Experimental Procedures

A. Washing of Amberlite XAD-2

Amberlite XAD-2 is washed as follows: 10 g Amberlite are stirred in 70 ml methanol for 15 min and washed on a Buchner-funnel with another 200 ml of methanol. The beads are then immediately washed with 1 l of water in order not to dry out. Then they are transfered to a column and washed slowly overnight with another 1 l of water. The beads are stored at 4°C in water containing 0.025% NaN$_3$ (Holloway 1973).

B. Assay for Acetylcholine Receptor

See p. 73.

C. Reconstitution Procedure

Lipid vesicles are prepared as follows: Egg lecithin (stock solution containing 10 mg lipid in 100 μl ethanol) is evaporated to dryness under a stream of nitrogen and then evacuated for 1 h. The lipid is dispersed in 5 ml ice-cold buffer solution of 10 mM Tris-HCl, pH 7.4, 200 mM KCl, 0.5% (w/v) Triton X-100. Amberlite XAD-2 (1.5 g wet weight, prepared as in A) is added and the solution is gently stirred at 4°C. After 30 min the receptor, pre-equilibrated in the same buffer, is added (preparation as described in Chapter 2.3). The lipid to protein ratio can be varied between 100:1 and 10:1 (w/w). The resulting solution is stirred for 24 h at 4°C and the resin is removed by filtration through glass wool. The recoveries of the different compounds used for reconstitution are given in Table 1.

Table 1. Recovery of toxin binding sites, phospholipid and ^{3}H-Triton X-100 in the phospholipid vesicles

Time [h]	α-bungarotoxin binding sites	Total inorganic phosphorus	^{3}H-Triton X-100
0	–	1.00 mg/ml (100%)	38.25 μmol (100%)
0.5	1.05 nmol (100%)	0.95 mg/ml (95%)	9.56 μmol (25%)
24	$>$0.84 nmol (80%)	0.80 mg/ml (80%)	$<$ 0.38 μmol ($<$1%)

The phospholipid vesicles were prepared as described in Section III. At time zero Amberlite XAD-2 was added to a mixture of phosphatidylcholine, ^{3}H-Triton X-100 and cold Triton X-100 (100% values). After 30 min purified acetylcholine receptor was added and recovery of the components was monitored

D. Characterization of the Resulting Structures

1. Sidedness

The sidedness determines the orientation of α-bungarotoxin binding sites, i.e., those facing the outside of the vesicles. Samples of incorporated acetylcholine receptor are assayed in the absence of Triton X-100 yielding the binding sites facing outward. The assay is repeated in presence of micellar Triton X-100 yielding total binding sites. In conditions of this experiment where small liposomes are formed, 85%–95% of the α-bungarotoxin binding sites of the reconstituted acetylcholine receptor are found to be orientated to the outside of the vesicles.

2. Density Gradients

Samples of (1) 0.1–0.2 ml of purified acetylcholine receptor and of (2) incorporated acetylcholine receptor are layered on the top of a 5%–30% continuous sucrose density gradient (total volume 12 ml) in a buffer of 10 mM Tris, pH 7.4, 0.2 M KCl) and centrifuged for 15 h at 200,000 g and 4°C. Fractions are then collected and assayed for α-bungarotoxin binding sites. Vesicular acetylcholine receptor float on the top of the gradient, whereas unincorporated protein will migrate into the gradient (Fig. 1).

3. Sepharose 4B Gel-Filtration

Sepharose 4B column chromatography allows the determination of the size distribution of the vesicles. The reconstituted material is passed through a column (2.5 × 30 cm) of Sepharose 4B preequilibrated with 20 mg of egg

H. Lüdi and U. Brodbeck

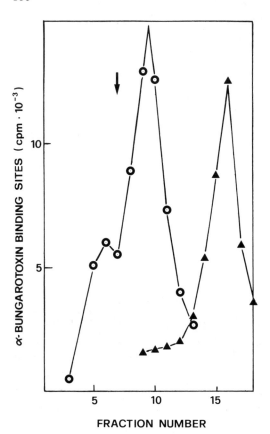

Fig. 1. Density gradient centrifugation of reconstituted and non-reconstituted pure acetylcholine-receptor. The sample (500 μl) was layered on top of a 5%–30% continuous sucrose density gradient prepared in a buffer of 10 mM Tris-HCl, pH 7.4 containing 0.2 M KCl. Centrifugation was carried out at 200,000 g for 15 h at 4°C. Fractions, 0.6 ml each, were collected from the bottom and analyzed for toxin binding activity (▲) Reconstituted and (○) non-reconstituted acetylcholine receptor. *Arrow* indicates peak fraction of catalase activity ($S_{4,w} = 11.4S$)

phosphatidylcholine (Hall and Brodbeck 1978). In order to calculate the ratio of remaining detergent molecules per lipid vesicle, reconstitution is carried out in presence of $[^3H]$-Triton X-100. Considering that a phospholipid vesicle of about 250 Å in diameter consist of about 5800 phosphatidylcholine molecules one molecule of Triton X-100 per 10 vesicles were found. Alternatively 1 molecule of detergent per 2.5 α-bungarotoxin binding sites could be calculated.

IV. Comments

The reconstitution procedure using Amberlite XAD-2 as detergent removing agent provides a reproducible and easy way to incorporate a membrane protein into a lipid bilayer structure. Besides acetylcholine receptor other proteins have been successfully reconstituted (Gerritsen et al. 1978; Chiesi et al. 1978). Depending on the nature of the protein used, freeze fracture electron microscopy reveals the presence of intramembraneous particles or not (Fig. 2). When the band 3 protein of human erythrocyte was reconstituted by the Amberlite XAD-2 method, particles could be visualized. On the other hand, glycophorin, although reconstituted, is not visualized as particles by freeze fraction electron microscopy (Lutz et al. 1979; Segrest et al. 1974). The results obtained with pure acetylcholine receptor give clear evidence that this protein is truly reconstituted into a lipid bilayer structure. Mere adsorption on the surface of the lipids by electrostatic interactions can be excluded by an experiment in which reconstitution is carried out in a medium of high ionic strength (in presence of 1 M NaCl) as similar results are obtained. In the Appendix a general strategy for the reconstitution of membrane proteins is detailed.

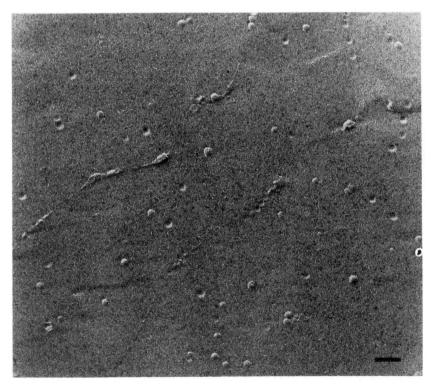

Fig. 2. Freeze fracture electron microscopy of reconstituted acetylcholine receptor. The *bar* represents 50 nm. Reconstitution was carried out at a protein to lipid ratio of 1 :10 (w/w). Picture courtesy of Dr. Ott

Appendix

The diagram below illustrates the general principal of development for a procedure to incorporate hydrophobic proteins into unilamellar phospholipid vesicles. Success of a reconstitution seems to be critically dependent of the appropriate lipid, nonionic detergent and hydrophobic resin to remove detergent with high selectivity.

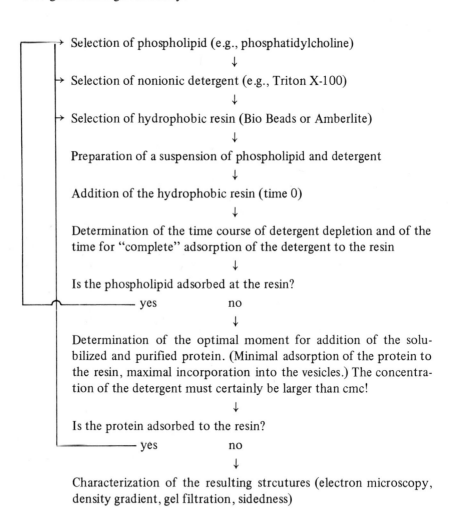

Selection of phospholipid (e.g., phosphatidylcholine)
↓
Selection of nonionic detergent (e.g., Triton X-100)
↓
Selection of hydrophobic resin (Bio Beads or Amberlite)
↓
Preparation of a suspension of phospholipid and detergent
↓
Addition of the hydrophobic resin (time 0)
↓
Determination of the time course of detergent depletion and of the time for "complete" adsorption of the detergent to the resin
↓
Is the phospholipid adsorbed at the resin?
———— yes   no
↓
Determination of the optimal moment for addition of the solubilized and purified protein. (Minimal adsorption of the protein to the resin, maximal incorporation into the vesicles.) The concentration of the detergent must certainly be larger than cmc!
↓
Is the protein adsorbed to the resin?
———— yes   no
↓
Characterization of the resulting strcutures (electron microscopy, density gradient, gel filtration, sidedness)

Acknowledgment. This work was supported by Swiss National Science Foundation grant No. 3.032-0.76

References

Brunner J, Skrabal P, Hauser H (1976) Biochem Biophys Acta 455:322–331

Changeux J-P, Heidmann T, Popot J-L, Sobel A (1979) FEBS Lett 105:181–187

Chiesi M, Peterson SW, Acuto O (1978) Arch Biochem Biophys 189:132–136

Eldefrawi ME, Eldefrawi AT, Mansour NA, Daly JW, Witkop B, Albuquerque EX (1978) Biochemistry 17:5474–5484

Elliot J, Raftery MA (1979) Biochemistry 18:1868–1874

Gerritsen WJ, Verkleij AJ, Zwaal RFA, van Deenen LLM (1978) Eur J Biochem 85: 255–261

Hall ER, Brodbeck U (1978) Eur J Biochem 89:159–167

Heidmann T, Changeux J-P (1978) Annu Rev Biochem 47:317–357

Holloway PW (1973) Anal Biochem 53:304–308

Howell J, Kemp G, Eldefrawi ME (1978) Tosteson DC, Ovchinnikov YuA. Latorre R (eds) Membrane transport processes, vol II. Raven Press, New York, pp 207–215

Huganier RL, Schell MA, Racker E (1979) FEBS Lett 108:155–160

Lutz HU, von Däniken A, Semenza G, Bächi Th (1979) Biochem Biophys Acta 552: 262–280

Schiebler W, Hucho F (1978) Eur J Biochem 85:55–63

Segrest JP, Gulik-Krzywicki T, Sardet C (1974) Proc Natl Acad Sci USA 71:3294– 3298

Wacker H, Müller F, Semenza G (1976) FEBS Lett 68:145–152

Wu WC-S, Raftery MA (1979) Biochem Biophys Res Commun 89:26–35

Reconstitution and Assay of Mitochondrial ADP/ATP Transport with a Partially Purified ADP/ATP Carrier Protein

G. BRANDOLIN, J. DOUSSIERE, G.J.M. LAUQUIN, and P.V. VIGNAIS

I. Introduction

ADP and ATP are transported across the inner mitochondrial membrane on a specific carrier, by an exchange-diffusion mechanism, with a 1:1 stoichiometry. The mitochondrial ADP/ATP carrier is selectively inhibited by a few natural toxic compounds, the atractylosides (atractyloside and carboxyatractyloside) and bongkrekic acid (Vignais 1976). The atractylosides do not readily penetrate the inner mitochondrial membrane (Vignais et al. 1973) whereas bongkrekic acid is a penetrant inhibitor at neutral or acidic pH (Kemp et al. 1971; Lauquin and Vignais 1976). On the basis of comparative inhibition studies of ADP/ATP transport in intact mitochondria and inside-out submitochondrial particles, it was concluded that the atractylosides attack the mitochondrial carrier from the outside of the inner mitochondrial membrane, and bongkrekic acid from the inside. Thus the above inhibitors are useful in reconstitution experiments to probe the asymmetrical insertion of the carrier protein into liposomes.

Procedures for purification of the ADP/ATP carrier protein and reconstitution of ADP/ATP transport have been recently reported (Shertzer et al. 1977; Kramer and Klingenberg 1979. Brandolin et al. 1980). The aim of the present text is to describe the preparation of a reconstituted ADP/ATP transport system by incorporation of the partially purified carrier protein into vesicles made from isolated phospholipids in suspension in glycerol solution and to present convenient kinetic assays of ADP/ATP transport with this system. The proteoliposomes are filled with ATP, and ADP/ATP transport can be followed by two different assays: (1) a radioactivity assay based on the uptake of $[^{14}C]ADP$ and (2) a luminescence assay that consists in monitoring the efflux of ATP dependent on uptake of external ADP.

II. Equipment and Reagents

A. Equipment

ultrasonic disintegrator (for ex. Branson W185D) fitted with a titanium
    microtip
fraction collector equiped with a UV photometer and a recorder
photometer linked to a recorder for detection of light emission in the firefly
    luciferase ATP assay. The Aminco Chem-Glow photometer, and the
    Aminco-Chance spectrophotometer have been satisfactorily used in our
    laboratory
scintillation counter
spectrophotometric 1-cm-path glass cuvette
Vortex mixer
25,000 g centrifuge
20 glass columns 0.6 X 40 cm
1 glass column 1.5 X 10 cm
a automatic adjustable pipette with disposable tips (200 μl)
microsyringes 5 μl, 10 μl, 50 μl
1- and 2-ml pipettes
nitrogen or argon tanks
stop watch

B. Reagents

egg phosphatidylethanolamine in chloroform (about 100 mg/ml) [purified
    following the method of Lea et al. (1955), and stored at $-20°C$ under
    argon, repurified before use]
beef heart cardiolipin in ethanol (about 5 mg/ml)
Dowex AG 1 X 8, 20/50 mesh, acetate form
hydroxyapatite for column chromatography, Biogel HTP (BioRad)
glycerol 136 mM — 1 liter
ADP 2.5 mM pH 7.4 — 2 ml
ATP 10 μM pH 7.4 — 2 ml
3-laurylamido-N,N'-dimethylpropylaminoxide (LAPAO) synthesized as
    described by Brandolin et al. (1980). If not available, LAPAO can be
    substituted by aminoxide WS35 which is a mixture of LAPAO (55%) and
    longer chain aminoxides (C_{14}, C_{16}, C_{18}). Aminoxide WS35 is obtained
    from Theo Goldschmidt AG, Essen, FRG
medium containing 20 mM ATP, 0.1 M glycerol, 0.1 mM EDTA, 10 mM
    tricine-KOH, final pH 7.4

medium containing 1 M Na_2SO_4, 0.1 M tricine-KOH, 1 mM EDTA, 10% LAPAO w/v, final pH 7.4 (this medium is kept at room temperature since it crystallizes in the cold)

medium containing 0.1 M Na_2SO_4, 0.1 mM EDTA, 0.5% LAPAO w/v, 10 mM tricine-KOH, final pH 7.4

medium containing 136 mM glycerol, 0.5 mM $MgSO_4$, 5 mM tricine-KOH, final pH 7.4

scintillation fluid. We routinely use a scintillation fluid containing: 24 g of 2,5 diphenyloxazole, 0.6 g of p-phenylene-2,2'-bis-methyl-4-phenyl-5-oxazole), 6 liters of toluene and 3 liters of Triton X-100; 2 ml of the radioactive liquid sample is added to 10 ml of scintillant (Patterson and Greene (1965)

tricine-KOH buffer 0.2 M, pH 7.4

carboxyatractyloside 0.2 mM (Boehringer)

bongkrekic acid 0.2 mM, obtained by the method of Lijmbach et al. (1970), modified by Lauquin et al. (1976) (aqueous solution neutralized by ammonia to pH 7)

$[^{14}C]ADP$ 5.10^6 dpm/μmol, 2.5 mM pH 7.4 (it is advisable to purify $[^{14}C]$ ADP before use by chromatography on a 0.5 $\times$ 2 cm column of Dowex AG 1 $\times$ 8, 200–400 mesh. After elimination of contaminating degradation components with 50 ml of distilled water, $[^{14}C]ADP$ is eluted with 2 N HCl

purified luciferase-luciferin. A premixed form of firefly luciferase and luciferin is commercially available as vials of lyophilized powder from Du Pont Inc., Wilmington, Delaware, USA, or from Lumac Systems AG, Basel, Switzerland (Lumit PM Kit)

beef heart mitochondria prepared by the method of Smith (1967), and suspended at a concentration of 50 mg protein/ml in 0.25 M sucrose, 10 mM Tris-HCl pH 7.4. The suspension of mitochondria can be kept frozen in liquid nitrogen and thawed just before use

III. Experimental Procedures

In reconstitution of the ADP/ATP transport activity, care must be taken to keep the period of time short between the isolation of the carrier protein and its incorporation into liposomes. It is advised to perform the different operations in the order described in the text.

A. Preparation of Liposomes

A mixture of purified phosphatidylethanolamine (36 mg in chloroform) and cardiolipin (4 mg in ethanol) is dried under argon at room temperature overnight to ensure complete evaporation of solvent. The dried lipid film is Vortex-dispersed in 1 ml of ice-cold medium made of 20 mM ATP, 100 mM glycerol, 0.1 mM EDTA, 10 mM tricine-KOH, final pH 7.4, and then submitted to sonication in an ice bath for 10 to 15 min under argon. The sonifier is equiped with a titanium microtip, using an output power of 40–50 W. Liposomes are cloudy to clear, depending on the phospholipid preparation. The liposome suspension can be kept in ice for a few hours before incorporation of the ADP/ATP carrier protein.

B. Preparation of Partially Purified ADP/ATP Carrier Protein and Incorporation into Liposomes

Hydroxyapatite is suspended in an ice-cold medium made of 0.1 M Na_2SO_4, 0.1 mM EDTA and 0.5% w/v LAPAO, 10 mM tricine-KOH, final pH 7.4. Care is taken to eliminate the fines. The gel suspension is poured in a 1.5 X 10 cm column to a height of 4 cm, and the column is stored at 2°C.

A 0.5 ml fraction of the stock suspension of beef heart mitochondria corresponding to 25 mg protein is mixed with 0.05 ml of the detergent medium made of 1 M Na_2SO_4, 0.1 M tricine-KOH, 1 mM EDTA and 10% LAPAO (w/v), final pH 7.4. After standing in ice for 5 min, the lyzed mitochondria are centrifuged at 20,000 g for 10 min. A 0.3 ml fraction of the supernatant fluid is layered on the hydroxyapatite column. Elution is performed at 2–4°C with the same LAPAO medium as that used for suspension of the hydroxyapatite. Elution of protein is monitored by UV absorbance. Fractions of 0.5 ml are collected. The first protein peak to appear consists of 70%–80% pure ADP/ATP carrier protein. Only the protein fraction corresponding to the top of the first elution peak is used for the reconstitution assay (0.5 to 1 mg protein per ml). 100 μl to 200μl of the carrier protein are added to 1 ml of the liposome suspension prepared as described above. After vigorous shaking, the mixture is incubated for 5 min in ice and is then sonicated in ice under argon for 15 to 20 s to facilitate incorporation of the carrier protein into liposomes. Proteoliposomes are freed of external ATP by passage at room temperature through a 0.6 X 25 cm column of Dowex AG 1 X 8, acetate form, 20–50 mesh, equilibrated with 136 mM glycerol; they are eluted just after the void volume (about 2.5 ml), and collected in a final volume of 5 to 10 ml. The pH of the proteoliposome suspension is adjusted to 7.4 by addition of tricine-KOH buffer (10 mM final concentration). These ATP-loaded proteoliposomes can be used for assay of ADP/ATP transport, immediately after preparation. They may also be used after one day aging at room temperature without severe loss of activity (only 10% to 20%).

C. Transport Assays

All assays are carried out at room temperature.

1. Radioactivity Assay

This assay is based on the uptake of [^{14}C]ADP into proteoliposomes. [^{14}C] ADP enters the proteoliposomes in exchange for the entrapped ATP.

Prepare 10 glass columns of 0.6 X 25 cm and fit their extremity with a plug of glass wool or any other porous support. Fill them with Dowex beads (AG 1 X 8, 20–50 mesh, acetate form) by aspiration of the bead suspension in 136 mM glycerol, the columns being held head downward. The height of the Dowex bed is adjusted to 25 cm, and the top of the bed is covered with a plug of glass wool. The columns are filled with 136 mM glycerol, till use for filtration of proteoliposomes.

Add 10 μl of 2.5 mM [^{14}C]ADP to 0.5 ml of the proteoliposome suspension and mix vigorously. Stop the [^{14}C]ADP uptake at intervals of time ranging between a few seconds and 30 min by addition of 15 μl of a mixture of 0.1 mM carboxyatractyloside and 0.1 mM bongkrekic acid. Let the suspension stand for 30 s with the inhibitors and then proceed with Dowex filtration.

For this, drain off the excess of glycerol in each column. Then fill the columns with 0.5 ml fractions of the proteoliposome suspension (having been incubated for different periods of time with [^{14}C]ADP). Elution is performed with the 136 mM glycerol solution. After removal of the dead volume (about 2.5 ml), collect the next 5 ml, mix, and use part of it (2 ml) for radioactivity counting.

The [^{14}C]ADP uptake, specific of ADP/ATP transport, has to be corrected for unspecific binding of [^{14}C]ADP (blank). Blanks are made by addition of 15 μl of the carboxyatractyloside-bongkrekic acid solution to proteoliposomes 30 s prior to [^{14}C]ADP, which corresponds to a final concentration of carboxyatractyloside and bongkrekic acid of about 3 μM. The incorporated radioactivity in the blanks is virtually the same whatever the length of the incubation period with [^{14}C]ADP. Routinely in our assays the [^{14}C] radioactivity incorporated in proteoliposomes in 10 s is at least twice

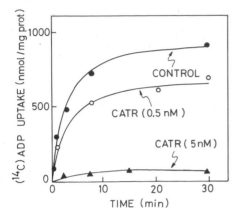

Fig. 1. Time course of ADP/ATP exchange assessed by radioactivity. The assay was performed as described in the text with an amount of proteoliposomes corresponding to 0.45 μg protein. The transport activity was corrected for blanks ([14C]ADP uptake insensitive to 3 μM carboxyatractyloside and bongkrekic acid). The partial inhibitory effect of two different concentrations (0.5 nM and 5 nM) of carboxyatractyloside (CATR) on the uptake of [14C]ADP is shown

that counted in the corresponding blank. Figure 1 shows a typical transport experiment with the inhibitory effect of low concentrations of carboxyatractyloside.

2. Luminescence Assay

This assay is based on the ADP-induced release of the ATP entrapped in proteoliposomes and on the continuous monitoring of the released ATP, using the bioluminescent assay. The luminescence medium is similar to that used for the radioactivity assay, except that it is supplemented with the luciferase preparation and Mg^{2+}. The luminescence emitted by a solution of purified luciferase-luciferin in presence of ATP can be measured with an ordinary photometer equiped with a sensitive photomultiplier. Characteristics of commercial instruments for luminescent assays have been detailed by Picciolo et al. (1978). In our laboratory, the measurements are performed with an Aminco-Chance dual beam spectrophotometer with the measuring light turned off. The 1-cm light path cuvette is covered with an aluminum foil on the three nonused faces to provide maximal light reflection and increase the sensitivity of the test.

Before embarking upon the luminescence assay of ADP/ATP transport, one has to verify the purity of the luciferase preparation and to set up conditions such that the luminescence is proportional to the ADP concentration. In our assays, we use a premixed form of firefly luciferase and luciferin commercially available as vials of lyophilized powder under the name of Lumit PM (Lumac Systems AG, Basel). The contents of one vial is dissolved in 3 ml of 136 mM glycerol and 5 mM tricine-KOH pH 7.4. When starting from separate components, it is advisable to use, as recommended by Lemasters and Hackenbrock (1980), luciferase at a concentration of 10 units/ml and luciferin at 7 μM.

a) Preliminary Tests

Assay for Contaminating Phosphokinases. Add in the photometer 0.5 ml of 136 mM glycerol, 0.5 mM $MgSO_4$ and 5 mM tricine-KOH pH 7.4, 0.05 ml of the luciferase-luciferin solution and 0.01 ml of 2.5 mM ADP. When the luciferase preparation is free from phosphokinase contaminants (e.g., adenylate kinase) capable of generating ATP from ADP, the recorded signal remains stable for minutes. Otherwise, the luminescence emission continuously increases. When the luminescence increase is too rapid, thus interfering with the ATP measurement, it is necessary to purify the luciferase preparation. We recommend filtration on Sephadex G200 equilibrated with 25 mM Hepes buffer pH 7.5 (Rasmussen et al. 1968). Other convenient methods of purification of luciferase are described by Rasmussen and Nielsen (1978). To the purified luciferase, luciferin has to be added.

Proportionality Assay. During the luciferase reaction, light emission is not always proportional to ATP concentration; this is in part due to product inhibition of the luciferase reaction (Lemasters and Hackenbrock 1980).

The following procedure is recommended for an approximate, but satisfactory proportionality test. The medium placed in the cuvette is the same as above, except that ADP addition is replaced by successive additions of 10 μl of 10 μM ATP to attain a luminescence signal of the same order as that expected for the ATP release in the ADP/ATP transport assay. The recorded light signals corresponding to equal additions of ATP must have the same size. If not, an increased amount of luciferase preparation is added. For exact measurement of ATP by luminescent assay, the reader is referred to the detailed articles by Lundin et al. (1976) and Lemasters and Hackenbrock (1980).

b) Luminescent Assay of ADP/ATP Transport

Fill the photometer cuvette with 0.5 ml of 136 mM glycerol, 0.5 mM $MgSO_4$ and 5 mM tricine-KOH pH 7.4, 0.05 ml of the luciferase solution and 0.05 ml of the proteoliposome suspension.

Initiate the ADP/ATP transport by addition of 10 μl of 2.5 mM ADP and record the ATP release by the luminescent assay. When the plateau is attained, calibrate the luminescence response by two successive additions of 10 μl ATP (Fig. 2). To better quantitative the luminescence signal, another calibration may be done at the end of the blank test (following paragraph).

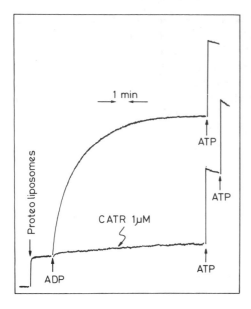

Fig. 2. Time course of ADP/ATP exchange assessed by luminescence. The assay was performed with 50 μl of proteoliposomes containing 0.40 μg added protein in a final volume of buffer of 0.5 ml as described in the text. ATP was released from the proteoliposomes upon addition of 50 μM ADP. The traces correspond to the continuous recording of ATP release, monitored by firefly luciferase bioluminescence, in absence or presence of 1 μM carboxyatractyloside (CATR) added 30 s before ADP. After the plateau was attained. an amount of 120 picomol of ATP in 10 μl was added to quantitate the luminescence signal

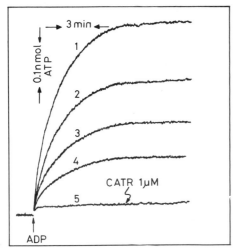

Fig. 3. Effect of proteoliposome concentration on kinetics of ADP/ATP transport. The transport activity was followed as in Fig. 2 by the luminescence assay, with various amounts of proteoliposomes containing 2.2 μg protein/ml; *1* 350 μl, *2* 250 μl, *3* 150 μl, *4* 100 μl. The sample incubated in the presence of 1 μM carboxyatractyloside (CATR) contained 350 μl proteoliposomes

Make a blank by adding 15 μl of 0.1 mM carboxyatractyloside and 0.1 mM bongkrekic acid to proteoliposomes. Leave the proteoliposome suspension in contact with the inhibitors for 30 s before addition of ADP.

Repeat the assay with increasing concentrations of proteoliposomes (Fig. 3) and check that both the rate and the total capacity of transport increase proportionally with the proteoliposome amount.

c) Suggested Experiments

Determine the K_M value for external ADP. For that, use the following final concentrations of added ADP: $2 - 5 - 10 - 15 - 20 - 30 - 40 - 50 -$ and 100 μM. In our hands, the K_M for ADP was in the range of 10 to 20 μM.

Test the effect of pH on the inhibition of ADT/ATP transport by bongkrekic acid. Proceed with a final concentration of ADP of 50 μM and a range of concentrations of bongkrekic acid between 5 nM and 2 μM, at pH 6.5 and 7.5. Remember that bongkrekic acid has to be protonated to penetrate the phospholipid bilayer (Kemp et al. 1971). The higher inhibition at pH 6.5 than at pH 7.5 suggests that the protonated form of bongkrekic acid attacks the ADP/ATP carrier protein from the inside of the phospholipid vesicles. This behavior is similar to that observed in mitochondria.

Compare the rates of ATP release (luminescence method) and [^{14}C]ADP uptake (radioactivity assay). The stoichiometry of ADP_{ex}/ATP_{in} exchange is 1:1.

Comments. 1. The [^{14}C]ADP uptake assay in the ADP/ATP exchange does not need Mg^{2+}. This is not the case for the luminescence assay of ATP release, since Mg^{2+} is required for the luciferase-luciferin reaction.

2. The luminescence assay of ADP/ATP transport makes it possible to continuously monitor the ADP-induced release of internal ATP and thus to study the initial kinetics of ADP/ATP exchange. The luminescence method is also more sensitive (at least ten times) than the radioactivity method.

3. Only a fraction of the purified ADP/ATP carrier protein added to liposomes behaves as competent carrier in proteoliposomes. Criteria for competent carrier estimation are: (1) high affinity binding for [^{3}H]atractyloside, (2) dose-response inhibition of transport by atractyloside or carboxyatractyloside (Brandolin et al. 1980). On the basis of these criteria, 3% to 6% of the added carrier protein is routinely found to be active in ADP/ATP transport in our experiments.

References

Brandolin G, Doussière J, Gulik A, Gulik-Krzywicki T, Lauquin GJM, Vignais PV (1980) Kinetic, binding and ultrastructural properties of the beef heart adenine nucleotide carrier protein after incorporation into phospholipid vesicles. Biochim Biophys Acta 592:592–614
Kemp A Jr, Souverijn JHM, Out TA (1971) Effect of bongkrekic acid and hydrobongkrekic acid on oxidation and adenine nucleotide transport in mitochondira. In: Quagliariello E, Papa S, Rossi CS (eds) Energy transduction in respiration and photosynthesis. Adriatica Editrice, Bari, Italy, pp 959–969

Kramer R, Klingenberg M (1979) Reconstitution of adenine nucleotide transport from beef heart mitochondria. Biochemistry 18:4209–4215

Lauquin GJM, Vignais PV (1976) Interaction of (^{3}H)bongkrekic acid with the mitochondrial adenine-nucleotide translocator. Biochemistry 15:2316–2322

Lea CH, Rhodes DN, Stoll RD (1955) Phospholipids: on the chromatographic separation of glycerophospholipids. Biochem J 60:353–363

Lemasters JJ, Hackenbrock CR (1980) The energized state of rat liver mitochondria. J Biol Chem 255:5674–5680

Lijmbach GWM, Cox MC, Berends W (1970) Elucidation of the chemical structure of bongkrekic acid. Isolation, purification and properties of bongkrekic acid. Tetrahedron 26:5993–5999

Lundin A, Rickardsson A, Thore A (1976) Continuous monitoring of ATP-converting reactions by purified firefly luciferase. Anal Biochem 75:611–620

Patterson MS, Greene RC (1965) Measurement of low energy beta-emitters in aqueous solution by liquid scintillation counting of emulsions. Anal Chem 37:854–857

Picciolo GL, Deming JW, Nibley DA, Chappelle EW (1978) Characteristics of commercial instruments and reagents for luminescent assays. Methods Enzymol 57:550–559

Rasmussen HN (1978) Preparation of partially purified firefly luciferase suitable for coupled assays. Methods Enzymol 57:28–36

Rasmussen H, Nielsen R (1968) An improved analysis of adenosine triphosphate by the luciferase method. Acta Chem Scand 22:1745–1756

Shertzer HG, Kanner B, Banerjee RK, Racker E (1977) Stimulation of adenine nucleotide translocation in reconstituted vesicles by phosphate and the phosphate transporter. Biochem Biophys Res Commun 75:779–784

Smith AL (1967) Preparation, properties and conditions for assay of mitochondria: slaughterhouse material, small-scale. Methods Enzymol 10:81–86

Vignais PV (1976) Molecular and physiological aspects of adenine nucleotide transport in mitochondria. Biochim Biophys Acta 456:1–38

Vignais PV, Vignais PM, Lauquin GJM, Morel F (1973) Binding of adenosine diphosphate and of antagonist ligands to the mitochondrial ADP carrier. Biochimie 55:763–778

Investigations on the Purple Membrane of Halobacterium halobium

K. SIGRIST-NELSON

I. Introduction and Aims

Halobacterium halobium is an extreme halophilic bacterium which synthesizes distinct purple-colored patches in its cell membrane when grown under conditions of light and low oxygen. The purple membrane, which can occupy up to 50% of the cell membrane area and is easily isolatable, functions as a light-driven proton pump. A single species of protein, bacteriorhodopsin, is present within the purple membrane. This protein, making up to 75% of the dry weight of the membrane, contains a single retinal moiety per protein molecule which is bound to the protein by a Schiff-base linkage to a lysine residue. Bacteriorhodopsin has been shown to function as a light-driven proton pump which creates an electrochemical gradient across the membrane. The electrochemical gradient can be used by the cells for ATP synthesis as described in Mitchells' chemiosmotic hypothesis.

Upon illumination of bacteriorhodopsin a cyclic sequence of absorption and conformation changes (the photochemical cycle, Fig. 1) is initiated. The life times of the intermediates are such that they are not detectable at room temperature by the usual spectroscopic methods. However, when bacteriorhodopsin (bR_{570}) is bleached by light in a concentrated salt solution

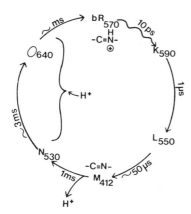

Fig. 1. Photoreaction cycle of bacteriorhodopsin. The *numbers* indicate the approximate half-times of intermediates

saturated with diethylether, an intermediate of the photo cycle, absorbing at 412 nm, can be measured. Ether serves to retard the 412 to 570 back reaction by a factor of 1000, so that the intermediate with the longest half-life accumulates and can be measured. Additionally, a release of protons concomitant with the formation of the 412 complex and an equal uptake on regeneration of the 570 complex occurs (Stoeckenius et al. 1979; Oester-helt and Hess 1973).

As mentioned, the purple membrane functions as an electrogenic hydro-gen ion pump driven by light. The orientation of bacteriorhodopsin is such that protons are pumped out of the cell, resulting in a net pH decrease in the external medium upon illumination. Sub-bacterial particles or cell envelopes may be prepared from intact bacteria (MacDonald and Lanyi 1975). One of the advantages of the envelope preparation is that pH gra-dients as well as membrane potential can be established or abolished at will. In these preparations bacteriorhodopsin is oriented inside-in, acidifying the suspending medium upon illumination. The pH changes observed are bipha-sic, as shown in Fig. 2. In contrast, the net direction of proton translocation of bacteriorhodopsin reconstituted in lipid vesicles is opposite to that found in whole cells. The reconstituted vesicles alkalinize the suspending medium upon illumination. Through the use of cardiolipin, a negatively charged phospholipid, and transient low pH during the reconstitution, it is possible, however, to achieve reconstitution of bacteriorhodopsin similar to the in vivo situation (Happe et al. 1977).

Furthermore, the proton-pumping activity of bacteriorhodopsin can be used to characterize properties of other co-incorporated membrane proteins. For example, the proton gradient created by illumination of reconstituted bacteriorhodopsin has been utilized to generate ATP formation by co-recon-stituted oligomycin-sensitive ATP'ase (Racker and Stoeckenius 1974). In the presented experiment the dicyclohexylcarbodiimide-binding proteolipid of the H^+-translocating ATP'ase is reconstituted with bacteriorhodopsin. Evidence has been presented that the DCCD-binding protein possesses

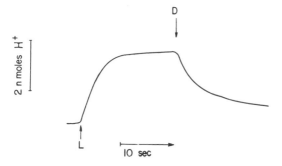

Fig. 2. Medium pH changes during illumination of bac-teriorhodopsin. *L* illumina-tion (light); *D* cessation of illumination (dark)

proton channeling activity which is inhibited upon addition of small amounts of DCCD (Nelson et al. 1977). Owing to its solubility in organic solvents, this small hydrophobic protein can be preferentially extracted from chloroplast membranes by n-butanol (Sigrist-Nelson et al. 1978). The butanol extract, containing the proteolipid and endogenous lipids is concentrated, butanol is removed, aqueous media added, and the mixture sonicated to form liposomes. Bacteriorhodopsin is co-sonicated with the proteolipid liposomes. The proton pumping activity of bacteriorhodopsin is then monitored in the presence and absence of dicyclohexylcarbodiimide. In the absence of DCCD, limited proton pumping activity is noted upon illumination. However, after preincubation of the liposomes with the reagent an increase in net proton pumping is noted due to inhibition of proteolipid-mediated proton translocation.

II. Equipment and Solutions

A. Equipment

centrifuge tubes
refrigerated centrifuge equivalent of Sorvall with SS-34 rotor
analytical centrifuge equivalent of Beckmann with 50 Ti, SW 27 or 40 rotors
recording spectrophotometer and cuvettes
density gradient mixer
glass electrode connected to a pH meter and recorder
thermostated vessel in which the glass electrode is inserted fitted with a
     magnetic stirring flea
magnetic stirrer
light source, for example a slide projector 150 watt
parafilm
probe type sonicator equivalent to Branson LS-75
bath type sonicator equivalent to Laboratories Supplies No. 611 2S P-IT
rotory evaporator with condensor attached to cryostat set at $-20°C$
high vaccum pump
Millipore filter apparatus with 0.45 μ HA filters

B. Solutions

basal medium adjusted to pH 7 containing:
     4.28   M NaCL
     0.081 M MgSO$_4$

0.010 M Na citrate
    0.027 M KCl
deoxyribonuclease
30%, 60% sucrose solutions
0.01 N NaOH and HCl
1 N HCl
4 M NaCl, saturated with ether
4 M NaCl, 0.05 M Tris-HCl pH 7
3 M KCl
0.15 M KCl
valinomycin 10 μg/ml in ethanol
carbonyl cyanide m-chlorophenyl-hydrazone (CCCP), 100 μM in ethanol
dicyclohexylcarbodiimide (DCCD), 40 mM in ethanol
n-butanol
chloroform:methanol (2:1 v/v)
asolectin (phospholipids isolated from soybean)
cardiolypin

III. Experimental Procedures

A. Isolation of the Purple Membrane of Halobacterium halobium

The bacteria are sedimented by centrifugation for 15 min at 16,000 g, 4°C. The sediment is suspended in basal medium and collected by centrifugation as above. The cell pellet is taken up in a small amount of water (protein concentration of approximately 50 mg/ml). The proton concentration is determined with help of an extinction coefficient (63,000 M^{-1} cm^{-1}) which is measured at 570 nm after short illumination with white light to transfer cis-retinal into trans-retinal. To apply the extinction coefficient for the all-transfiguration the measurement should be carried out immediately following illumination. Deoxyribonuclease (0.25 mg per 50 mg protein) is added. Cell membranes are collected by centrifugation for 30 min at 100,000 g, 4°C. The pellet is taken up in a small amount of water. In the next step the various membranes of the lysed bacteria are separated by density gradient centrifugation. A 30%–60% (w/w) sucrose gradient is used. The gradient is formed by mixing 18 ml 30% sucrose with 25 ml 60% sucrose with a gradient mixer. The lysate membranes are carefully added to the top of the gradient and the samples are centrifuged for 20 h, 15°C at 100,000 g in a swing-out rotor. The violet band (purple membrane) is carefully siphoned off with a Pasteur pipette, diluted with water and sedimented by centrifugation (30 min, 4°C, 100,000 g) (Osterhelt and Stoeckenius 1973).

Reversible Bleaching of Bacteriorhodopsin by Light in the Presence of Ether.
The purple membrane preparation is added to 4 M NaCl saturated with ether.
An aliquot of the purple membrane is added to a water-jacketed chamber
maintained at 20°C. Small amounts of 0.01 N NaOH or HCl are added to
achieve a final pH of approximately 6. The chamber is closed with parafilm.
After 5 min in the dark (dim room light) the sample is illuminated. The pH
changes in the purple membrane suspension are monitored by a glass electrode
connected to a pH meter with a recorder. The dark-light cycle is repeated.
Upon exposure to light the characteristic purple color of the membrane
changes to yellow due to accumulation of the 412 nm complex in the presence
of ether. Accompanying the "backing up" of the 412 nm complex is acidifica-
tion of the medium resulting from stoichiometric proton release during the
photocycle. The return of the purple color is noted with reformation of the
570 nm complex and accompanying medium alkalinity (a proton being
taken up). The same experiment is repeated with purple membranes which
have not been exposed to ether. No change in medium pH should be noted
with the dark-light sequence. Minor differences in medium pH are sometimes noted due to the effect of the light source on the pH electrode.

B. Isolation of Cell Envelopes

Cells (450 ml) in the early stationary phase (A_{660} = 1.7) are harvested by
centrifugation for 20 min at 10,000 g, 4°C. The cells are suspended in basal
salt medium, approximately 1/3 to 1/2 the volume of the growth medium,
and are recovered by centrifugation. The cells are further washed with
1/3 volume 4 M NaCl, 0.05 Tris HCl pH 7 and resuspended with a syringe in
1/30 volume (15 ml). The cooled preparation is sonicated four times for
15 s with a probe-type sonicator (Branson LS-75). Deoxyribonuclease
0.2 mg per ml solution is added. The lysate is then diluted fivefold with
buffered NaCl. Residual cells are removed by centrifugation at 4500 g for
20 min, 4°C. The supernatant is removed and the envelopes recovered by
centrifugation at 46,000 g for 30 min, 4°C. The pellet is suspended in
unbuffered 4 M NaCl, recovered by centrifugation and finally suspended in
4 M NaCl at a protein concentration of 1 to 2 mg/ml. The envelopes are
stored at 4°C (MacDonald and Lanyi 1975).

C. Demonstration of Light-Driven Proton Pumping Activity in:

1. Intact Cells

Cell suspension (1 ml) is added to a water-jacketed glass chamber at 25°C
containing 1.8 ml 4 M NaCl, 0.2 ml 3 M KCl. 1 N HCl (5 μl) is added to

achieve a final pH of approximately 6.The chamber is flushed with nitrogen and closed with parafilm. pH changes are monitored as in A with a glass electrode-pH meter and recorder combination. After the pH value of the medium is stable the cell suspension is illuminated with light for about 5 min until the pH under illuminating conditions has reached a constant level. With discontinuation of the illumination the return of the medium pH to its original level is followed.

2. Cell Envelopes

The resealed cell envelope preparation obtained in B is utilized for an experiment similar to C.1. An aliquot of cell envelopes is added to the water-jacketed chamber containing 3 M KCl. Light-dependent proton pumping is measured as in C.1.

3. Reconstituted Bacteriorhodopsin

Inside-Out Orientation. 5 mg asolectin is added to a small amount of chloroform:methanol (2:1). The organic solvent is evaporated under nitrogen. 1 ml of 0.15 M KCl is added and the suspension is vortexed vigorously. The vesicle preparation is sonicated in a bath-type sonicator, under nitrogen atmosphere, till clarity. Bacteriorhodopsin is added to an aliquot of the sonicated vesicles and the preparation is further sonicated. An aliquot of the bacteriorhodopsin liposomes is added to the water-jacketed glass chamber containing 0.15 M KCl. The liposome preparation is illuminated and light-dependent proton pumping is measured by a pH electrode-pH meter-recorder as in C.1 and C.2. Additionally 5 μl valinomycin, a potassium ionophore, is added to the glass chamber and proton pumping activity is monitored. In the presence of valinomycin the net proton pumping observed is elevated due to the ionophores charge neutralization effect on the liposomal system. In a separate experiment the proton ionophore CCCP (5 μl) is added to bacteriorhodopsin liposomes. In contrast to valinomycin, addition of CCCP abolished all observed pumping activity due to its permeability-enhancing properties for protons.

Inside-In Orientation. Cardiolypin, 1 mg/ml (5 ml) is sonicated in a bath-type sonicator under a N_2 atmosphere in 0.15 M KCl at 30°C till a fairly clear solution is obtained. 1 ml of the sonicated liposomes is adjusted to pH 2.5 with 0.1 N HCl while another ml of liposomes is left as is (approximately pH 4.2). Purple membrane suspension is added to both prepared liposomes. The mixture is sonicated for a *very* short period of time, 10 to

20 s, with a probe-type sonicator, following which the pH is *immediately* raised to around 6. Great care must be taken in the sonication step because of the tendency of bacteriorhodopsin to bleach under acid conditions. Proton pumping is then assayed.

D. Bacteriorhodopsin Reconstituted with the DCCD-Binding Proteolipid

Chloroplast membranes are concentrated by centrifugation (20,000 g for 10 min, 4°C) and are taken up in distilled water at a concentration of 15 mg protein/ml. Protein is determined by the Lowry procedure in the presence of 0.1% sodium dodecyl sulfate after precipitation of the chloroplast membrane protein by 80% acetone. Eight ml of chloroplast membranes are injected into 160 ml n-butanol at 4°C under vigorous stirring for 10 min. The butanol is then centrifuged at 4°C, 12,000 g for 10 min. Following centrifugation the butanol is filtered through a Millipore filter to remove remaining precipitated protein. The filtered butanol is concentrated till dryness by a rotory evaporator under high vacuum with the condensor cooled to -20°C. During the evaporation the flask containing the butanol is maintained at 20°C. When dry, 6 ml n-butanol is added to the flask to dissolve the dried extract which is then transferred in 2 ml portions to glass tubes. The butanol is removed by evaporation under nitrogen and 0.5 ml 0.15 M NaCl is added per tube. The mixture is flushed with nitrogen, sealed with parafilm and sonicated in a bath-type sonicator at 20°C twice for 15 min. Bacteriorhodopsin (0.1 ml, 3 mg protein/ml) is added and the mixture is sonicated for a further 15 min.

Light-induced proton pumping is assayed as described previously. NaCl (0.15 M, 2 ml) is added to the thermostated assay chamber. The bacteriorhodopsin-proteolipid liposomes (0.1 ml) are added and the light-induced increase in the external pH is measured. Dicyclohexylcarbodiimide is added (40 nmol) and incubated for 30 min with the liposomes. The proton pumping activity is again measured and compared to the earlier measurement. DCCD is additionally added to bacteriorhodopsin liposomes not containing proteolipid, in which case no effect on the proton pumping activity should be noted.

References

Happe M, Teather R, Overath P, Knobling A, Oesterhelt D (1977) Direction of proton translocation in proteoliposomes formed from purple membrane and acidic lipids depends on the pH during reconstitution. Biochim Biophys Acta 465:415–420

Lowry OH, Rosebrough NJ, Farr AL, Randall RJ (1951) Protein measurement with the folin-phenol reagent. J Biol Chem 193:265–275

MacDonald R, Lanyi J (1975) Light-induced leucine transport in *Halobacterium halobium* envelope vesicles: A chemiosmotic system. Biochemistry 14:2882–2889

Nelson N, Eytan E, Notsani B, Sigrist H, Sigrist-Nelson K, Gitler C (1977) Isolation of a chloroplast proteolipid and demonstration of dicyclohexylcarbodiimide-sensitive proton translocation. Proc Natl Acad Sci USA 74:2374–2378

Oesterhelt D, Hess B (1973) Reversible photolysis of the purple complex in the purple membrane of *Halobacterium halobium*. Eur J Biochem 37:316–326

Oesterhelt D, Stoeckenius W (1973) Isolation of the cell membrane of *Halobacterium halobium* and its fractionation into red and purple membranes. Methods Enzymol 31:667–678

Racker E, Stoeckenius W (1974) Reconstitution of purple membrane vesicles catalyzing light-driven proton uptake and adenosine triphosphate formation. J Biol Chem 249:662–663

Sigrist-Nelson K, Sigrist H, Azzi A (1978) Characterization of the dicyclohexylcarbodiimide-binding protein isolated from chloroplast membranes. Eur J Biochem 92:9–14

Stoeckenius W, Lozier R, Bogomolni R (1979) Bacteriorhodopsin and the purple membrane of halobacteria. Biochim Biophys Acta 50:215–278

Enzymology of Succinate:Ubiquinone Reductase in Detergent Solution and Reconstituted Membranes

P. WINGFIELD and H. WEISS

I. Introduction and Aims

Succinate:ubiquinone reductase, or electron transfer Complex II (for a comprehensive treatment see Fleischer and Packer 1978), is an enzyme of the inner mitochondrial membrane and cytalyses the following reaction:

Succinate + ubiquinone → fumarate + dihydroubiquinone

The enzyme will serve as an example of a membrane enzyme which utilizes a hydrophobic substrate.

A. Assay of the Enzyme Activity in Detergent Solution

Succinate:ubiquinone reductase has been isolated from *Neurospora* mitochondria as a monodisperse protein-detergent complex (Weiss and Kolb 1979). A micelle of the nonionic detergent Triton X-100 [aklyl-phenyl poloxyethylene ether, tert. $C_8 E_{9.6}$, see Helenius et al. (1979)] replaces the phospholipid bilayer. It is assumed that the detergent is bound to that section of the enzyme which was originally embedded in the membrane.

In the mitochondrial inner membrane the enzyme and the hydrophobic substrate ubiquinone-10 are located within a continuous lipid bilayer. Thus, enzyme-substrate interactions can occur by lateral translations in the plane of the membrane. In detergent solution the enzyme and the hydrophobic substrate are inserted into discrete micells. Since the total micelle concentration is much greater than the concentration of the enzyme-bound micelles, most of the hydrophobic substrate is partitioned into free micelles. Hence, the enzyme and substrate are in a discontinuous phase with respect to one another. For an enzymic reaction to occur, the hydrophobic substrate must transfer from free micelles to the enzmyme-bound micelles according to the following reaction scheme:

$$E.M + M.Q \rightleftharpoons E.M.Q + M \tag{1}$$
$$E.M.Q + S \rightleftharpoons E.M.QH_2 + F \tag{2}$$

Where E is the enzyme succinate:ubiquinone reductase; M a detergent micelle; Q is ubiquinone-10; QH_2 is dihydroubiquinone-10, S is succinate and F is fumarate. Reaction (1) is the substrate transfer reaction, and reaction (2) the enzymic reaction. The rate of the substrate transfer reaction can be rate-limiting, depending on the type of nonionic detergent used (Weiss and Wingfield 1979).

By coupling the enzymic reaction (2) to an auxiliary nonenzymic reaction (3), which recycles dihydroubiquinone to ubiquinone within the enzyme-bound micelle, the kinetics of the substrate transfer reaction can be ignored. A coupled assay is represented by the following reaction scheme:

$$E.M + M.Q \quad \rightleftharpoons E.M.Q + M \tag{1}$$
$$E.M.Q + S \quad \rightleftharpoons E.M.QH_2 + F \tag{2}$$
$$E.M.QH_2 + D \rightarrow E.M.Q + DH_2 \tag{3}$$

Where, in addition to the symbols used above, D is dichloroindolphenol and DH_2 is reduced dichloroindolphenol. Under the conditions of the coupled enzyme assay, the kinetics of the intrinsic enzymic reaction are measured. This will be demonstrated in the experimental section.

The coupled assay can also be used to measure the equilibrium position of the substrate transfer reaction. The principle involved is as follows. Because the auxiliary reaction (3) is faster than the enzymic reaction (2), the substrate concentration in the enzyme-bound micelles will have a constant value. The substrate concentration is controlled only by the equilibrium position of the substrate transfer reaction (1). If we assume that ubiquinone-10 equilibrates freely among all micelles, free and enzyme-bound, the equilibrium constant of the transfer reaction will be equal to 1 according to Eq. (1):

$$\frac{[E.M.Q] \cdot [M]}{[E.M] \cdot [M.Q]} = 1 \tag{1}$$

When the ubiquinone concentration is equal to half the total micelle concentration according to Eq. (2):

$$[Q_{total}] = 1/2 \ [M_{total}] \tag{2}$$

then half of the micelles will contain a substrate molecule according to Eq. (3):

$$[M.Q] = [M] \tag{3}$$

Under the conditions of Eq. (1) and (3), it follows that half of the enzyme-bound micelles will contain one substrate molecule according to Eq. (4):

$$[E.M] = [E.M.Q] = 1/2 [E_{total}] \tag{4}$$

Thus, the ubiquinone concentration required for half maximal activity (equivalent to the app K_m) will be half the total micelle concentration independent of the actual micelle concentration according to Eq. (5):

$$app K_m = 0.5 [M] \tag{5}$$

This derivation will be experimentally verified in the practical section.

B. Assay of the Enzyme Activity in Reconstituted Membranes

Isolated succinate:ubiquinone reductase can be reincorporated into phospholipid bilayers by the removal of Triton X-100 from a mixture of the enzyme-Triton X-100 complex and mixed micelles of Triton X-100 and phospholipid. The membrane bound enzyme can be assayed in aqueous solution only when ubiquinone has been incorporated into the membrane. This can be done in an effective and nondisrupting manner by the calcium induced fusion of reconstituted succinate:ubiquinone reductase with liposomes containing ubiquinone-10. This is schematically represented as follows:

$$E_{membr.} + Q_{membr.} \rightarrow E.Q_{membr.} \tag{1}$$
$$E.Q_{membr.} + S \rightleftharpoons E.Q.H_{2\ membr.} + F \tag{2}$$
$$E.Q.H_{2\ membr.} + D \rightarrow E.Q_{membr.} + DH_2 \tag{3}$$

Where $E_{membr.}$ is the enzyme succinate : ubiquinone reductase incorporated into a phospholipid bilayer; $Q_{membr.}$ is ubiquinone-10 containing liposomes; S is succinate; F is fumarate; D is dichloroindolphenol and DH_2 is reduced dichloroindolphenol. Reaction (1) represents the membrane fusion reaction — a continuous lipid phase is produced between enzyme and ubiquinone-10. Reaction (2) represents the enzymic reaction — ubiquinone is reduced by succinate and reaction (3) the auxiliary redox reaction which recycles dihydroubiquinone-10 to ubiquinone-10 within the membrane. This reaction also serves as indicator reaction.

II. Equipment and Solutions

A. Equipment

double-wavelength spectrophotometer with thermostated cell holders
    (e.g., Perkin Elmer 156 Double-Wavelength Sepctrophotometer)
disposable plastic 1 cm cuvettes for spectroscopy

microsyringes (10, 25, 50, and 100 μl)
micropipettes (10, 20, 50 μl)
magnetic stirrer
stoppered test tubes
nitrogen gas source
sucrose density gradient marker
swing out rotor (e.g., Beckman SW40)
density gradient fractionator (e.g., MSE)
water bath

B. Solutions and Materials

Triton X-100, 10% in water
dichloroindolphenol, 0.1% in water
Na-succinate, 1.0 M in water
Tris-acetate buffer, 1.0 M, pH 7.0
ubiquinone-10, 5 mg/ml in ethanol
butylated hydroxytoluene (BHT), 10 mg/ml in ethanol
phenylmethylsulfonylfluoride (PMSF), 1.0 M in ethyl acetate
sucrose, 10% w/v and 40% w/v containing 50 mM Tris-acetate, pH 7.0,
    2 mM Na-succinate, 0.1 mM PMSF and 5 μM BHT
Bio Beads SM2 (Bio Rad Laboratories), washed as previously described
    (Holloway 1973)
CaCl$_2$, 50 mM in water
EDTA, 100 mM in water, pH 7.0
soybean phosphatidyl choline (Carl Roth)
bovine brain phosphatidyl serine (Sigma)

III. Experimental Procedures

Succinate-ubiquinone reductase was isolated as a enzyme-detergent complex
from *Neurospora crassa* mitochondria as previously described (Weiss and
Kolb 1979).

A. Enzymology of Succinate:Ubiquinone Reductase in Detergent Solution

The assay medium is composed of 50 mM Tris-acetate pH 7.0, 5 mM Na-
succinate, 50 μM dichloroindolphenol, 2–10 μM ubiquinone-10 and 0.1%–
1.0% Triton X-100. The enzymic reaction is started by the addition of 5 μl

of a 5 μM enzyme solution [the molarity is based on one cytochrome b per mol of enzyme. The molecular weight of the enzyme is about 120,000 (Weiss and Kolb 1979)]. The reduction of dichloroindolphenol is monitored with the wavelength pair 610 and 750 nm using $\epsilon_{red-ox} = 20$ mM^{-1} cm^{-1}. The volume of the assay is 2 ml and the temperature 30°C. The maximal activities (V^Q_{max}) and the apparent Michaelis constants (app K^Q_M) are derived from double reciprocal plots of reaction rates versus ubiquinone concentration at 0.1, 0.2, 0.5 and 1% detergent concentrations. The results will show that V^Q_{max} is not affected by the concentration of detergent used. However, the app K^Q_M will increase with increase in the detergent concentration. To determine the equilibrium position of the substrate transfer reaction (1), the app $K_M^Q/[M]$ is plotted versus [M]. [M] is the concentration of Triton X-100 micelles and is calculated by the following formula:

$$\frac{(\text{total detergent concentration in } \% - \text{critical micellar concentration in } \%) \ 0.10}{\text{average molecular weight of the micelle}} = \text{molar micelle concentration}$$

The critical micellar concentration of Triton X-100 is 0.3 mM, the average molecular weight of the micelle is 90,000 (Helenius et al. 1979). See Weiss and Winfield (1979) for illustrations of the above data plots.

B. Incorporation of Succinate–Ubiquinone Reductase into Phospholipid Bilayers

Twenty mg phosphatidyl choline and 5 mg phosphatidyl serine, are dissolved in 3 ml chloroform and dried with a stream of nitrogen at 40°C. The residue is twice dissolved in 3 ml ether and dried with nitrogen. 5 ml of a solution of 50 mM Tris-acetate pH 7.0, 0.5% Triton X-100, 5 μM butylated hydroxytoltolnene is added to the phospholipid film. The mixture is sonicated until a clear solution is obtained, usually 2–3 min, and then the solution is centrifuged at 100,000 g for 15 min to sediment undispersed phospholipid and pieces of titanium from the sonicator tip.

   Equal volumes of approx. 2.5 mg/ml succinate:ubiquinone reductase solution and the phospholipid solution which is about 5 mg/ml, are mixed. To remove Triton X-100, 200 mg/ml Bio Beads SM are added to the combined solution which is then gently stirred at 4°C for 2.5 h. The Bio Beads are removed by filtration through glass wool packed in a Pasteur pipette. Samples of 1.5–2.5 ml are applied to 12 ml 10%–40% (w/v) sucrose density gradients which contain 50 mM Tris-acetate pH 7, and centrifuged at 280,000 g for 18 h at 4°C. Fractions are displaced from the gradients using

60% (w/v) sucrose and a gradient fractionator. The succinate:ubiquinone reductase membranes band isopynically at a density of $1.11-1.13$ g/cc^3 and the protein to lipid ratio of the membranes is approximately 0.90. Any excess phospholipid remains at the top of the gradients.

C. Preparation of Ubiquinone-10 Containing Liposomes

Sixteen mg phosphatidyl choline and 4 mg phosphatidyl serine are mixed with 1 mg ubiquinone-10 and dried from chloroform and ether as described above. Four ml of 50 mM Tris-acetate pH 7.0, containing 5 μM BHT is added, and the mixture is sonicated for $10-15$ min keeping the glass sonication vial in stirring ice cold water. The pH is checked after 5 min, and if necessary adjusted to pH 7.0. The solution is centrifuged at 1,000,000 g for 15 min at 4°C and any pelleted material is discarded. At least 90% of the phospholipid remains in solution. The solution is stored in the dark at 4°C.

Using this method unilamellar vesicles of approx. 0.02 μm in diameter are formed in which 5% by weight of the phospholipid is contributed by ubiquinone-10. This ubiquinone-10 to phospholipid ratio is similar to that found in the mitochondrial inner membrane.

D. Assay of Membrane-Bound Succinate:Ubiquinone Reductase

The membrane-bound succinate:ubiquinone reductase ($2-5$ μM based on the cytochrome b content) is preincubated with 20 mM sodium succinate at room temperature for 20 min. This incubation activates the enzyme (Fleischer and Packer 1978). $1-25$ μl samples of the enzyme membranes are mixed with $0-100$ μl samples of the ubiquinone-10 containing liposomes. The mixtures are incubated firstly with 5 mM CaCl$_2$ for 15 min and secondly with 10 mM EDTA for 5 min, both incubations are at 30°C. The mixtures are assayed for succinate:ubiquinone reductase activity in a medium composed of 50 mM Tris-acetate pH 7.0, 10% sucrose, 5 mM Na-succinate and 34 μM dichloroindolphenol. The assay is started by the addition of the membranes. The reduction of dichloroindolphenol is monitored as described above. The total assay volume is 2 ml and the temperature 30°C. The V^Q_{max} and app. K^Q_M are derived from double reciprocal plots of reaction rates versus the concentration of liposomal ubiquinone. The results will show that the maximal activity (V^Q_{max}) obtained for the membrane-bound enzyme is similar to that determined for the enzyme in detergent solution.

References

Fleischer S, Packer L (eds) (1978) Methods in enzymology, vol 53. Biomembranes, biological oxidation, mitochondrial and microbial systems. Academic Press, London New York

Helenius A, Casli DRM, Fries E, Tanford C (1979) Methods Enzymol 56:734—749

Holloway PS (9173) Anal Biochem 53:304—308

Weiss H, Kolb HJ (1979) Eur J Biochem 99:139—149

Weiss H, Wingfield PT (1979) Eur J Biochem 99:151—160

Isolation and Reconstitution of Band 3, the Anion Transporter of the Human Red Blood Cell Membrane

Z.I. CABANTCHIK

I. Introduction and Aims

The assignment of a given membrane function to a particular class of membrane polypeptides rests on their identification by chemical labeling methods and on their isolation in a reasonably pure form by means conservative of the function. Solubilization and reconstitution of membrane functions is important for assessing the modulatory role of lipids and other protein constuents as well as for gaining further insights into the architecture of functional sites and the underlying molecular mechanisms (Racker 1977).

Membrane transport systems are specialized functional entities which expedite geographic changes of solutes across membranes, usually by chemically and electrically silent mechanisms. Most of these systems have no intrinsic markers such as prosthetic groups or enzymatic activities which can be used to identify them in the course of membrane disruption, solubilization, and protein fractionation. Therefore, appropriate procedures have to be developed, first for their chemical identification, and second for their isolation and reconstitution into closed membrane systems amenable for assaying their transport properties (e.g., planar membranes, closed vesicles, or intact cells).

A. The Anion Transport System of Human Red Blood Cells

For various reasons we have chosen the anion transport system of human red blood cells to exemplify the methodology of functional isolation and reconstitution of membrane transport proteins. First, an extensive series of chemical labeling studies (Cabanthchik et al. 1978; Passow et al. 1977; Knauf 1979) has already indicated the involvement of a particular class of 95,000 molecular weight polypeptides (according to SDS-gel electrophoresis) in $Cl\text{-}HCO_3^-$ transmembrane exchange (Cabantchik et al. 1978; Passow et al. 1977) These polypeptides migrate to an area in the gel known as band 3 (B 3) (Steck 1978). Physiologically, the exchange subserves the requirement for efficient transport of CO_2 from tissues to lungs, which is otherwise limited by the poor plasma solubility of CO_2. Mechanistically, CO_2 produced in tissues, diffuses into red cells as they pass through blood capillaries and is quickly converted to HCO_3^- by intraerythrocytic carbonic anhydrase. This HCO_3^- exchanges with plasma Cl^- across the membrane by an electronuetral process mediated by the anion transporter, so that within a second (the mean passage time of erythrocytes through capillaries) all anions are close to reaching electrochemical equilibrium (Cabantchik et al. 1978; Knauf 1979). We can safely say that in analogy with the red cell specialization for O_2 binding which stems from the high cytoplasmic content of hemoglobin, the presence of an efficient and specialized anion exchange mechanism reflects the high content of band 3 in erythrocyte membrane (25% dry weight of all membrane polypeptides or 12% of the membrane dry weight).

Second, the relative ease and high yield with which human erythrocyte membranes can be prepared and processed has made this system one of the best-studied animal transport systems both in terms of transport kinetics (Deuticke 1977; Cabantchik et al. 1978; Knauf 1979), of chemical structure and of membrane arrangement of B3 (Guidotti 1980; Tanner 1978). Furthermore, various procedures for isolating B3 at different degrees of purity have been described (Tanner 1979).

B. Band 3 (B3), the Anion Transporter

B3 is a glycoprotein with a total carbohydrate content estimated variously as 3%–15% by weight and a hydrophobic index of 37.5% (Steck 1978; Tanner 1979). As might be expected for a protein involved in transport, it is intimately associated with the membrane. It can be dissolved out of the membrane by detergents but not by ionic stresses (e.g., extraction with high or low salt concentration, high pH etc.) or protein perturbants (e.g., urea, lithium-3.5-diodosalycilate, etc.) which strip off the majority of membrane polypeptides (Marchesi et al. 1976; Steck 1978). The protein which is composed of a single polypeptide chain is arranged asymmetrically in the membrane, with all the carbohydrate facing outward and all the sites for binding glycolytic enzymes (Steck 1978) and hemoglobin (Salhanay et al. 1980) facing inward (the N-terminus is also located on the inner surface). The polypeptides reside in the membrane possibly as homologous dimers (Steck 1978; Guidotti 1979) transversing the membrane several times (Tanner 1978; Guidotti 1980). Since the protein as a whole does not rotate in a plane perpendicular to the membrane surface (Cherry et al. 1976; Cherry and Nigg 1980) it is likely that transport occurs over a restricted domain of the polypeptide. As recent studies have shown (Rothstein et al. 1980) two neighboring transmembrane integral segments which comprise less than 15%

of the polypeptide mass may contain the sites for specific nonpenetrating inhibitors which are probably also the sites of transport. In intact membranes, these sites are accessible to the aforementioned inhibitors (e.g., DIDS or DNDS) only from the outer surface (Passow et al. 1977; Barzilay and Cabantchik 1979).

C. Isolation and Reconstitution of B3 (Rationale and Aims)

Three basic strategies developed for the isolation and reconstitution of B3 will be presented. The common element of these strategies rests on the fact that each of them provides an experimental object suitable for testing anion transport activity (either sulfate or chloride) in a "closed" membrane system such as a vesicle or a cell. This is accomplished either by isolating B3 in its natural lipid milieu (A), by extracting B3 with nonionic detergents and reconstituting it with externally added lipids (B), or by implanting the isolated B3 into plasma membranes of living cells lacking Cl-transfer capacity. We shall refer to procedures (A) as negative purification (NP) and to procedure (B) as solubilization and reconstitution (SR). These two are designated in vitro procedures in contrast to (C) which can be viewed as an in vivo reconstitution (IR) procedure (a schematic view of the various procedures is portrayed in Fig. 1).

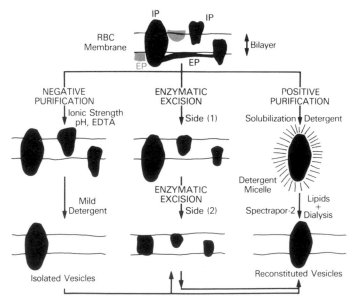

Fig. 1. Schematic diagram describing isolation of B3, VG, and VGB3 by various procedures. *IP* and *EP* represent intrinsic and extrinsic proteins, respectively

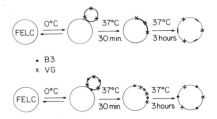

Fig. 2. Implantation of either VG or VGB3 into Friend erythroleukemia cells (FELC)

In order to provide a firmer basis for the putative role of B3 in anion transport, several isolation and reconstitution procedures had to be developed (Volsky et al. 1979). It was found that B3 isolated by NP or incorporated into liposomes of defined lipid composition (SR) expressed only partial capacity to transfer the analog anion sulfate. However, B3 implanted into plasma membranes of living cells (lacking Cl-transfer properties) (IR) displayed Cl-transport properties commensurate with those observable in intact erythrocytes (Volsky et al. 1979). The method of in vivo reconstitution makes use of the fact that solubilized and reconstituted Sendai virus envelopes have demonstrable capacity to fuse with and integrate in plasma membrane of living eukaryotic cells (Volsky and Loyter 1978b). The envelopes can be coreconstituted with membrane proteins and thus be used as vehicles for inserting these proteins into recipient cells (Volsky et al. 1979) (see Fig. 2). Thus the method provides a useful tool for assaying isolated components of membrane transport systems, particularly in those cases in which in vitro reconstitution is not easily attainable, either because of lipid composition unfavorable for reconstitution, leakiness of vesicles, or simply because of technical limitations in assaying rapid transport processes across systems of large surface/volume properties (e.g., liposomes). The rationale of this work rests on the assumption that plasma membranes of intact cells can provide to exogenous proteins a matrix suitable for transport measurements and an appropriate environment to favorable express their putative functions.

II. Equipment and Solutions

A. Equipment and Accessories

1. In Vitro Procedures (NP and SR)

Sorvall Refrigerated Centrifuge RC2B or equivalent
Spinco Ultracentrifuge or equivalent
Hematocrit centrifuge

bath sonifier
rotaevaporator (Büchi or equivalent)
N_2-gas container with pressure regulator
spectrophotometer
magnetic stirrers
vacuum line for aspirating supernates
stop watches
thermostated baths
automatic pipettes
disposable syringes 2 ml (or mini columns)
rack (for disposable syringes) which can be kept at ice-cold temperature
spectrophotometric glass cuvettes (3 ml)
liquid scintillation counter
vials for liquid scintillation counter
spectrapor-2 dialysis bags (1 cm diameter)
centrifuge tubes polycarbonate (40 ml)
centrifuge tubes polyallomer or polycarbonate for ultracentrifuge (10—15 ml)

2. In Vivo Procedures (IR)

Same as (1) but also:
thermostated circulating unit for low temperatures
thermostated vessels with round-type stirrer
hood for sterile work
inverted microscope
glass syringes with Luer-Lock ends
Sartorius filter holders
Sartorius or Millipore filters (1.2 μm pore size)
equipment for culture growing of cells in suspension
fluid dispenser for scintillation fluids
equipment for sterile work:
disposable petri dishes and centrifuge tubes
assorted graduated pipettes
assorted Pasteur pipettes
assorted glassware

B. Solutions

PBS (phosphate buffered saline) NaH_2PO_4 20 mM, NaCl 140 mM pH 8.0
    (titrated with 1 N NaOH)
PB (phosphate buffer) NaH_2PO_4 20 mM, pH 8.0 (titrated with 1 N NaOH)

PB 1/5 Dilute PB 1:5 with H_2O

PBS 1/5 Dilute PBS 1:5 with H_2O

Triton X-100 0.045% in PBS 1/5 (containing dithiothreitol 0.5 mM). This solution was deoxygenated by bubbling N_2 at 0°C prior to the experiment

PBS 1/15 Dilute PBS 1:15 with H_2O. Deoxygenate with N_2

TBS (tris-Cl buffered saline) Tris-Cl 50 mM, NaCl 100 mM, pH 7.4

PMSF (Sigma) Phenylmethylsulfonylfluoride 0.20 mM in PBS 1/15 or TBS

PBS 1/15 NaN_3 0.1 mM containing defatted bovine serum albumin (BSA) 10 mg/ml

TCS (tricine saline) Tricine NaOH 20 mM, NaCl mM, pH 7.4

EDTA 0.5 M, pH 7.4

HSS HEPES-sulfate, HEPES 20 mM, Na_2SO_4 50 mM, sucrose 20 mM, pH 7.0 (titrated with NaOH) (deoxygenated with N_2)

HSS 1/1 Dilute HSS 1:1 with H_2O (deoxygenated with N_2)

SM Sucrose 220 mM, $MnSO_4$ 0.2 mM

HSSM Same as HSS but containing 0.2 mM $MnSO_4$

Bradford reagent (Bio Rad) for protein measurements or Lowry reagent for protein measurements

Dowex AG 1 X 8 50–100 mesh (Bio Rad) succinate form, rinsed with 10 mg/ml bovine serum albumin (fatty acid free) in Na-succinate 60 mM, sucrose 20 mM, pH 7.4. Keep at 0°C

[^{3}H] Triton X-100 (NEN)

NBD-albumin (nitrobenzyldiazole conjugated albumin) (10:1 molar ratio) 500 μg/ml HSS

[^{3}H] H_2DIDS (4,4′-diisothiocyano-2,2′-ditritiostilbene disulfonic acid) (0.25 Ci/nmol) 1 mM in PBS (Nuclear Research Center − Negev, Israel)

DNDS (4,4-dinitro2,2′-stilbenedisulfonic acid) 10 mM in HSS and 0.5 mM in 350 mM sucrose

Sephadex G-25 Fine

Na-succinate 80 mM pH 7.4

Na-citrate 80 mM pH 7.4

scintillation fluid (Aquasol, Liquifluor, Triton-based fluid or equivalent)

TCS as above but sterilized

DMEM Dulbecco's Modified Eagle's medium (sterilized)

FCS fetal calf serum (Gibco) (sterilized)

DM Dulbecco's medium (sterilized)

$CaCl_2$ 0.1 M (sterilized)

1 unit of whole human blood (ca. 400 ml)

Friend erythroleukemia cells (ca. 10^8 cells) grown in Dulbecco's Modified Eagle's Medium (DMEM)

Sendai virus harvested and purifeid as described by Volsky and Loyter (1978b)

III. Experimental Procedures

The experiments are designed so that each group will carry out only one section (either IIIA, IIIB or IIIC + D). Steps III-1 and III-2 are common to all groups.

1. Red Blood Cells

Take whole human blood or packed cells (recently outdated) and add 20 ml to each of 8 centrifuge tubes (polycarbonate, transparent). Centrifuge on a Sorvall RC2B centrifuge (bring up to 5000 rpm and stop). Aspirate supernatant and remove buffy coat overlaying red cells. Add PBS, pH 8.0 to cells, cover the tube with parafilm and mix the suspension. Remove the parafilm and centrifuge as above. Repeat all the steps until the supernatant is clear and no visible buffy coat remains as a red cell overlay.

1a. Labeling of Cells Either with $[^3H]_2$ DIDS (see Sect. IIIE) or with DIDS (see Sect. IIIF)

2. Red Blood Cell Ghosts

Collect all cells (ca. 50 ml), take 40 ml packed cells and cool to $0°C$. Add all the cells at once to a 500 ml Erlenmeyer containing 280 ml pB, pH 8.0 and placed in an ice-cold bath with vigorous stirring for about 10 min. Then distribute the suspension to 8 centrifuge tubes (same as above) and centrifuge ($0°C$) at 15,000 rpm for 8 min. Place tubes in ice-bucket, remove the hemolysate by aspiration and make sure to leave the fluffy sediment of red pink ghosts. Remove the ghosts from the bottom by tilting the centrifuge tube. Transfer the ghosts to a different centrifuge tube (this step is necessary in order to free the ghosts from a sticky pellet which contains proteases. Whenever such a pellet is identified transfer the ghosts to a different tube). Add PB, PH 8.0 ($0°C$) to each tube, mix the suspension and centrifuge as above. Repeat these steps twice. Then resuspend the ghosts in a 1:15 dilution of PB, pH 8.0 (PB 1/5) $0°C$, centrifuge, aspirate and repeat the washing once more. The ghosts should look white (opalescent texture). Collect the packed ghosts in one tube and measure their volume. Keep them at $0°C$.

A. Isolation of NP Membranes (Wolosin et al. 1977; Cabantchik et al. 1978) (see Fig. 3)

1. Removal of Extrinsic Proteins (High Ionic Strength)

Wash the ghosts with 5 volumes of PBS, pH 8.0, 0°C, leave for 20 min at 0°C and then centrifuge at 15,000 rpm for 8 min (0°C). Remove supernatant and collect residue.

2. Removal of Sialoglycoproteins

To each vol. of the residue add 6 vol. of Triton X-100 0.045% in PBS 1/5, pH 8.0 (0°C) and leave for 20 min at 0°C. Centrifuge at 18,000 rpm × 10 min (0°C). Repeat the detergent extraction once more with 3 vol. of Triton, this time add [^{3}H] Triton X-100 (5 × 10^6 cpm/ml final). Centrifuge and remove supernatant.

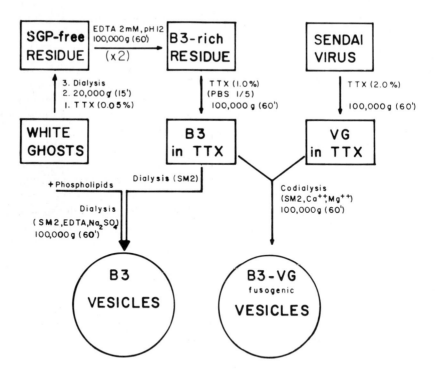

Fig. 3. Schematic description of B3 isolation by negative and positive purification and by enzymatic excision. *IP* and *EP* intrinsic protein and extrinsic proteins, respectively

3. Removal of Triton X-100

Resuspend the Triton residue in 5 vol. of PBS 1/15 (0°C) and centrifuge as in (4). Combine all the residues. Take 5 μl sample for [^{3}H] measurements. Add 2 vol. PMSF (0.15 mM in PBS 1/15) in order to neutralize possible traces of residual protease activity. Leave for 10 min at 0°C and place all the residue in a Spectrapor-2 dialysis bag. Immerse the bag in 100 vol. of PBS 1/15, albumin 10 mg/ml, NaN$_3$ 0.1 mM (bubble N$_2$). Take samples of 1 ml for [^{3}H] counting (ca. every 8 h) until [^{3}H] in dialysate reaches a constant value (ca. 24 h). Replace the external abumin NaN$_3$-PBS 1/15 medium (10 vol.) and make sure no trace of [^{3}H] emerges in the dialysate during the next 8 h. Remove the content of the bag and sample the [^{3}H]. Centrifuge at 18,000 rpm for 10 min, 0°C.

4. Removal of Strongly Associated – Extrinsic Polypeptides (NP-Membranes)

To each vol. of Triton-residue (end of step 3) add 5 vol. of 2 mM EDTA, pH 12, vortex thoroughly and incubate for 10 min at 0°C (N$_2$!). Centrifuge at 100,000 g for 30 min in a polyallomer[1] centrifuge tube. Remove the supernatant and repeat once again the extraction. Wash the final pellet with 5 vol. of Hepes-sulfate-sucrose (HSS) medium pH 7.2, 0°C, and centrifuge as above. Take up the pellet in 5 vol. of HSS (0°C) and determine on an aliquot the protein content either by the Lowry-SDS method or the Bradford method. Assume ca. 4 mg/ml protein in this preparation of NP membranes. Freeze the unused portion of NP at N$_2$ temperature and keep at − 70°C.

B. Isolation of B3 by SR (Volsky et al. 1979; Wolosin 1980) (see Fig. 3)

1. Proceed up to the end of step A3 as described for NP membranes (III A3) but do not use [^{3}H] Triton at this step.

2. Solubilization of B3 and Dialysis of Detergent

To each 4 mg proteins of sialoglycoprotein free membranes (end of step A2) add 10 mg of Triton X-100/0.5 ml TBS pH 7.4 containing 0.2 mM PMSF and 25 μCi of [^{3}H] Triton. Incubate at room temperature for 30 min and

---

1 Polyethylene or polypropylene tubes are also acceptable, polycarbonate deteriorates in alkaline media

centrifuge at 100,000 g for 1 h. Transfer the supernate to a spectrapor dialysis bag after taking a 10 μl aliquot for [^{3}H] counting and dialyze 60 h at 4°C against 100 vol. of HSS containing 2 mg/ml albumin (to sequester the detergent) 0.5 mM NaN$_3$ and 0.10 mM PMSF. Every 8 h check the [^{3}H] that either remained in the bag or [^{3}H] that egressed out. Change the dialysis solution.

3. Reconstitution of B3 with Externally Added Phospholipids

To the solubilized B3 partially free of Triton X-100 (ca. 0.4% after 60 h) add first n-octylglucoside (20 mg detergent per mg protein or 5% final concentration) and subsequently transfer the above suspensions into a test tube containing N$_2$-dried phospholipids (synthetic DL-lecithin: Phosphatidyl serine 94:6, w/w) such that the final protein-phospholipod ratio is 1:4, w/w. Vortex the suspension and dialyze it further against 100 vol. of HSS 1/1 containing 5 mM EDTA, 0.5 mM NaN$_3$, and 2 mg/ml albumin. Remove the bag content and sample for protein (Lowry SDS), for [^{3}H]-Triton, for SDS-acrylymide gel electrophoresis and for organic phosphate (Wolosin et al. 1977). Freeze the final SR preparation with liquid N$_2$ and keep at -70°C.

C. Coreconstitution of B3 with Envelope Glycoprotein of Sendai Virus for Obtaining Fusogenic Vesicles (Volsky et al. 1979; Cabantchik and Loyter 1979) (see Fig. 3)

1. The glycoproteins of Sendai virus envelopes were obtained by solubilization of 20 mg of intact virons in 2 ml TBS containing 40 mg Triton X-100 (20 min at room temperature) and sedimentation at 100,000 g for 1 h to remove the nucleocapsid components. The resulting supernate was composed mainly of two envelope glycoproteins, HN (hemagglutinin-neuraminidase) and F (fusion factor) at a concentration of 2 mg protein per ml. The supernate was treated for 20 min at room temperature with 0.1 mM PMSF. Reconstituted viral proteins (VG) were obtained by Spectrapor-2 dialysis (0°C, 75 h) of the solubilized envelopes against a pH 7.4 buffer containing 10 mM Tris-HCl, 0.5 mM NaN$_3$, 2 mM CaCl$_2$, 2 mM MgSO$_4$, 0.15 mM PMSF and 2 mg/ml albumin. The content of the dialyzed material was subsequently centrifuged at 100,000 g for 45 min to yield a VG pellet of ca. 1.5 mg reconstituted viral glycoproteins.

2. Solubilized B3 (500 μg protein in 0.3 ml) prepared as described earlier (III B, step 2, undialyzed) was added to 0.7 ml (1.00 mg protein) solubilized

VG (III C step 1, undialyzed). The mixture was codialyzed for 75 h and sedimented as described above for VG. The pellet containing, among others, vesicles of hybrid nature (VGB3) yielded ca. 700 μg protein (see Volsky et al. 1979 for further details). Freeze the two preparations VG and VGB3 with liquid N_2 and keep at $-70°C$.

D. Implantation of B3 into Friend Erythroleukemia Cells (FELC) by Viral Mediated Fusion of VPB3 Vesicles (see Fig. 2)

FELC grown in DMEM containing 10% calf or newborn serum were centrifuged for 1000 rpm X 4 min and washed once with warm TCS. Fusion of either VG of VGB3 with cells was performed by addition of reconstituted viral vesicles (20–30 μg protein in 20 μl) to 10 ml of cells in TCS containing 5 mM $CaCl_2$ at 4°C. After 10 min at 4°C during which agglutination occurs, the cells were centrifuged, washed twice with the above TCS buffer (0°C) and subsequently incubated with the same buffer for another 20 min at 37°C to promote vesicle-cell fusion. The cells were washed once again with TCS alone and finally they were resuspended in growth medium for 3 h. Implantation yields of B3 transfer into cells are ca. 30% of total added band 3.

E. Estimation of Isolation and Reconstitution Yields Using $[^3H]H_2$ DIDS as a Specific Marker of B3

Red blood cells (10 ml) washed with PBS (III-1) at 25% hematocrit were reacted with $[^3H]H_2$ DIDS (0.5 Ci/mmol) (50 μM final concentration) for 30 min at 37°C with shaking. The cells were centrifuged and washed twice with PBS containing 5 mg/ml BSA and twice with PBS. Ghosts prepared from $[^3H]H_2$ DIDS labeled cells as shown in III-2, gave 1 mol $[^3H]H_2$ DIDS per mol of ghost-B3 (assuming 6.6×10^{-10} mg protein/ghost and 1.2×10^6 B3 polypeptides/ghost (Steck 1978). This $[^3H]$-labeled-B3 can be used in parallel experiments to estimate the yield of the various purification and reconstitution procedures. These estimations gave the following figures: 30% and 20% of the original (red cell) B3 for NP and SR, respectively, and 6% of the red cell B3 for the overall IR procedures (i.e., isolation, reconstitution, and implantation).

F. Inactivation of B3 Transport Capacity by Reaction of Cells with DIDS

Red blood cells (10 ml) washed with PBS (III-2) at 25% hematocrit suspension were reacted with DIDS (50 μM final concentration) for 20 min at 37°C with shaking. The cells were washed, centrifuged, and processed for ghost preparation and B3 isolation as shown in previous sections.

G. Measurements of Sulfate-Exchange in Isolated Vesicles (NP) or Reconstituted Vesicles (SR)

1. Sealing of NP Vesicles and Loading of [^{35}S]SO$_4$

Thaw the NP preparation (III AG). Centrifuge at 50,000 g × 30 min and discard the supernate. Make up a 2 mg/ml protein suspension in HSS + MnSO$_4$ + 5 × 10^8 cpm/ml [^{35}S]Na$_2$SO$_4$ + 0.5 mM NBD-albumin. Bubble N$_2$ into the suspension and sonicate for 10 min at 0°C in a bath sonicator. Transfer the tubes containing sealed vesicles to 37°C and incubate for an additional hour. (Note: if required, it is possible to incorporate into vesicles DNDS, a reversible nonpenetrating inhibitor of anion transport at ca. 0.2 mM prior to sonication.)

2. Sealing of SR Vesicles and Loading of [^{35}S]SO$_4$

Thaw the SR preparation (III-B3) and freeze (with liquid N$_2$) and thaw it two more times in a medium containing HSS + 5 × 10^8 cpm/ml [^{35}S]Na$_2$SO$_4$ + 0.5 mM NBD albumin + 0.2 mM DNDS. Sonicate as shown above (III-F1).

3. Removal of Extravesicular NBD-Albumin (all steps are carried out at 4°C) (NP of SR)

Take a 10 ml disposable syringe and fill it up with a Sephadex G-25 fine suspension in H$_2$O. Centrifuge the syringe into a centrifuge tube in a swing-out centrifuge (3000 rpm × 10 min) at 0°C to remove the H$_2$O. Carefully lay over the vesicle suspension and centrifuge as above for 5 min. Collect the vesicles free of extravesicular NBD-albumin. This step might need to be repeated twice for complete removal of external NBD-albumin.

4. Removal of Extravesicular [^{35}S]SO$_4$ (NP of SR)

Transfer the vesicles to 0°C and pour the suspension over a Dowex AG 1 × 8 50–100 mesh column (0.5 cm diameter × 2 cm length) (succinate or citrate form) kept at $0°C$. Rinse the column with 1.0 ml SM. Using a UV lamp detect and collect the NBD-albumin containing vesicles ($0°C$) (green yellow fluorescence). Stop rinsing the column when drops become weakly fluorescent (ca. between 0.7–1.3 ml). Keep the vesicles at $0°C$.

5. [^{35}S] Sulfate Efflex Measurements (NP or SR)

Prior to step 4 prepare 50 mini columns of Dowex AG 1 × 8 50–100 mesh citrate or succinate form (0.5 cm diameter × 1 cm length) and keep at $0°C$. Bring the extra-resin fluid to resin level. To initiate fluxes jet 100 μl of NP into 1 ml HSS media preincubated to indicated temperatures (1 to 6). At indicated times, withdraw 80 μl aliquot and lay over a single column ($0°C$) (this stops the flux and sequesters the extravesicular [^{35}S]). Rinse with 1.0 ml SM ($0°C$) and collect the eluate.

Systems	1	2	3	4	5	6
Medium	HSS	HSS + DNDS (0.2 mM)	HSS	HSS	HSS	HSS
	$30°C$	$30°C$	$25°C$	$15°C$	$5°C$	$37°C$
Sampling times	15″	15″	15″	15″	15″	15″
	30″	3′	3′	5′	10′	30″
	1′	10′	4′	10′	20′	1′
	2′	10′	10′	10′	30′	2′
	5′	60′	20′	30′	60′	5′
	10′		60′	60′		10′
	15′					60′

Note: After withdrawing the various aliquots transfer the tube containing the remaining vesicles to $37°C$ for full isotopic equilibration. Take 0.5 ml samples for [^{35}S] counting in either Aquasol or Triton X based scintillation fluid. Take also 80 μl aliquot of the 1–6 suspension of [^{35}S] for counting (add 420 μl medium in order to bring all samples to equal volume).

6. Analysis of Data (Sulfate Fluxes)

a) For each system, plot on a linear paper cpm (t)/cpm (0′) versus t where cpm (t) are the counts at time t cpm (0′) are the counts at time 0, hence the radioactivity of 80 μl suspension. Verify that the efflux profile is exponential. Ideally, if all the vesicles have functional protein, the intravesicular [^{35}S] should approach zero at time ∞ (say at 1 h 37°C). However, in practice up to 30% of the [^{35}S] might be entrapped in nontransporting vesicles. The total radioactivity at t = 1 h (37°C) is designated cpm (∞). Extrapolate the linear part of the profile to obtain the Y-intercept. This value represents the experimental cpm (0) which might differ by a few % from the above ideal cpm (0′).

b) Since the above represents a typical isotope dilution experiment, it can be represented by

$$\ln \left[\frac{cpm\ (t) - cpm\ (\infty)}{cpm\ (0) - cpm\ (\infty)} \right] = -kt$$

where k is the rate constant of efflux.

Plot the expression in parenthesis versus t on semilogarithmic paper and obtain the slope −k. Alternatively, calculate the half-time of efflux t 1/2, that is the value of t when the expression in parenthesis is equal to 1/2. In this case:

$$t\ 1/2 = \frac{\ln 2}{k} \quad \text{or}\ k = \frac{0.693}{t\ 1/2}$$

Calculate k from t 1/2.

c) In order to obtain the energy of activation Ea, plot on a semilogarithmic paper k versus 1/T (where T is the absolute temperature). Since $\ln k = A - \dfrac{Ea}{RT}$, Ea can be calculated from the slope of the experimental line.

d) In order to calculate the transport efficiency of B3 in NP, compare the k in vesicles (kv) with the k in ghosts (kg). Assume the following: (a) The density of B3 in the membranes of NP and ghosts is the same. (b) The radius of NP is 0.14 μm; the radius of ghosts is 6.32 μm. (c) t 1/2 at 25°C is 60 min. Since the rate constants are directly affected by the surface (S) to volume (V) ratio (S)/(V), it is expected that for fully functional B3 in:

$$\frac{kv}{kg} = \frac{[(S)/(V)]v}{[(S)/(V)]g} \qquad S = 4\pi r^2$$

$$\qquad\qquad\qquad\qquad\qquad V = \frac{4}{3}\pi r^3$$

$$\frac{kv}{kg} = \frac{rg}{rv} \qquad\qquad \frac{S}{V} = \frac{3}{r}$$

$$\frac{kv}{kg} = 22$$

From the computed value of kv (25°C) and the above value of kg calculate the efficiency of the system in NP.

e) Similar computations can be done for SR in which case $\dfrac{kv}{kg} = 40$. Assume also that the average density of B3 proteins/phospholipid is roughly the same as in ghosts or in NP (i.2., 1:4). The latter can and should be estimated from the protein content and phospholipid content of the various preparations.

f) Summarize the k values of sulfate efflux at 30°C from the two types of vesicles in the various media (± DNDS). Based on the assumption that DNDS inhibits anion transport only exofacially (i.e., on the external surface of B3 in intact cells) try to assess the disposition of B3 in NP vesicles and in SR vesicles.

g) From the computed Ea of sulfate efflux in NP and SR vesicles and from the knowledge that the corresponding Ea in intact cells is ca. 30 Kcal/mol, assess the role of the membrane matrix in determining Ea of a trans-membrane phenomenon mediated by an integral membrane protein.

H. Measurements of Cl-Exchange in FELC Reconstituted with B3 by the IR Method

1. The Experiment

Four classes of FELC will be used for this experiment. Untreated FELC (control), FELC fused with VG and FELC fused either with VGB3 or fused with VPB3 prepared from B3 which was inactivated with DIDS (III-F).

Following the 3-h cultivation period (IIIP) the various FELC (each at 10^8 cells in 10 ml growth medium) were centrifuged at 1000 rpm $\times$ 4 min, the supernate discarded and the cells resuspended in 0.8 ml growth medium containing Na$[^{36}Cl]$ (6×10^7 cpm/ml). The cell suspensions were incubated for 30 min at 37°C with gentle shaking. At the end of this period the whole suspension was gently overlaid a centrifuge tube containing 30 ml of PBS ± 2 mg BSA + 2 mM DNDS at 0°C, quickly centrifuged for 1/2 min at 5000 rpm. The supernate was discarded and the cell pellet was quickly transferred into a 0°C thermostated vessel with constant stirring and with circulation jacket containing 30 ml PBS + 2 mg/ml BSA. Upon addition of cells to medium efflux of $[^{36}Cl]$ from cells is initated. Samples of 1–2 ml are taken by suction with a 5 ml syringe connected to a Sartorius filter holder using a prefilter and a 1.2 μm filter. These filtration units mounted with the various filters are kept at -10°C until used. Samples are taken at the indicated times and the filtrates are transferred from the syringe to a test tube. At the end of 30$'$ the remaining contents are transferred to a test tube and incubated for an hour at 37°C with shaking.

1a) At the end of this hour an ∞ (infinity) sample is taken by filtration as above.

Systems (sampling times)

FELC	FELC + VP	FELC + VPB3	FELC + VPB3-DIDS
20″	20″	20″	20″
40″	40″	40″	40″
60″	60″	60″	60″
90″	90″	90″	90″
120″	120″	120″	120″
3′	3′	3′	3′
5′	5′	5′	5′
10′	10′	10′	10′
20′	20′	20′	20′
30′	30′	30′	30′
∞ (30′, 37°)	∞ (30′, 37°C)	∞ 30′, 37°C)	∞ (30′, 37°C)

For counting the [^{36}Cl] transfer 0.4 ml of each filtrate into a scintillation vial, containing 3.5 ml of scintillation fluid.

2. Analysis of Data

The calculation of the rate constant of Cl-efflux is essentially the same as that shown for vesicles (III G6) except that here we shall use the following equation:

$$\ln \left[\frac{cpm\ (\infty) - cpm\ (t)}{cpm\ (\infty) - cpm\ (0')} \right] = -kt$$

$$\ln [y] = -kt$$

Here y represents the fraction of radioactivity left in cells at each time t. Plot cpm (t) versus t and obtain by extrapolation cmp (0′). Then use the above equation to calculate y and plot y (t) versus t on a semilogarithmic paper. Calculate k for each system.

One optional system which can be tested in order to gain information on the orientation of B3 implanted into FELC (FELC-VgB3) is that of Cl fluxes in FELC-VGB3 in a PBS medium containing 1 mM DNDS.

3. From the Cl-efflux profiles of the different systems try to ascertain how many exponential components are required in order to describe the above flux, particularly in FELC-VGB3 system. Compare the FELC-VGB3 (DIDS) system with that of FELC-VGB3 measured in the presence of external DNDS and assess the disposition of the implanted B3. (Note: the k for Cl-exchange in intact red blood cells at 0°C, pH 7.2 is ca. 15″).

Acknowledgment. This project was based on work supported by the USA-Israel Binational Science Foundation (Jerusalem) and by NIH (USA) grant No. GM 27753.

References

Barzilay M, Cabantchik ZI (1979a) Anion transport in red blood cells. II. Kinetics of reversible inhibition by nitroaromatic sulfonic acids. Membrane Biochem 2:255—281

Barzilay M, Cabantchik ZI (1979b) Anion transport in red blood cells. III. Sites and sidedness of inhibition by high affinity reversible binding probes. Membrane Biochem 2:282—300

Cabantchik ZI, Loyter A (1980) Functional characterization of isolated membrane transport systems, the erythrocyte anion transporter as a model. In: Lassen UV, Ussing HH, Wieth JO (eds) Membrane transport in erythrocytes. Alfred Benzon Symp 14. Munksgaard, Copenhagen, pp 373—388

Cabantchik ZI, Knauf PA, Rothstein A (1978) The anion transport system of the red blood cell: The role of membrane protein evaluated by the use of "probes". Biochim Biophys Acta (Membrane Dev) 515:239—302

Cherry RJ, Nigg EA (1980) Molecular interactions involving band 3: information from rotational diffusion measurements. In: Lassen UV, Ussing HH, Wieth JO (eds) Membrane transport in erythrocytes. Alfred Benzon Symp 14. Munksgaard, Copenhagen, pp 130—142

Cherry RJ, Burkli A, Busslinger M, Schneider F, Parish GR (1976) Rotational diffusion of band 3 proteins in the human erythrocyte membrane. Nature (London) 263: 389—393

Deuticke B (1977) Properties and structural basis of simple diffusion pathways in the erythrocyte membrane. Rev Physiol Biochem Pharmacol 78:1—97

Guidotti G (1980) The structure of the band 3 polypeptide. In: Lassen UV, Ussing HH, Wieth JO (eds) Membrane transport in erythrocytes. Alfred Benzon Symp 14. Munksgaard, Copenhagen, pp 300—311

Knauf PA (1979) Erythrocyte anion exchange and the band 3 protein. In: Bronner F, Kleinzeller A (eds) Current topics in membranes and transport, vol 12. Academic Press, London New York, pp 251—363

Marchesi VT, Furthmayr H, Tomita M (1976) The red cell membrane. Annu Rev Biochem 45:667—698

Passow H, Fasold H, Lepke S, Pring M, Schuhmann B (1977) Chemical and enzymic modification of membrane proteins and anion transport in human red blood cells. In: Miller MW, Shamoo AE (eds) Membrane toxicity. Plenum Press, New York, pp 353—377

Racker E (1977) Perspectives and limitations of resolution-reconstitution experiments. J Supramol Struct 6:219—228

Rothstein A, Ramjeesingh M, Grinstein S (1980) The arrangement of transport and inhibitory sites in band 3 protein. In: Lassen UV, Ussing HH, With JO (eds) Membrane transport in erythrocytes. Alfred Benzon Symp 14. Munksgaard, Copenhagen, pp 330—344

Salhanay JM, Cordes KA, Gaines ED (1980) Light scattering measurements of hemoglobin to the erythrocyte membrane. Evidence for transmembrane effects related to a disulfonic stilbene binding to band 3. Biochemistry 19:1447—1454

Steck TL (1978) The band 3 protein of the human red cell membrane. A review. J Supramol 8:311–324

Tanner MJA (1978) Erythrocyte glycoproteins. In: Bronner F, Kleinzeller A (eds) Current topics in membranes and transport, vol 11. Academic Press, London New York, pp 279–325

Tanner MJA (1979) Isolation of integral membrane proteins and criteria for identifying carrier proteins. In: Bronner F, Kleinzeller A (eds) Current topics in membranes and transport, vol 12. Academic Press, London New York, pp 1–51

Volsky J, Loyter A (1978a) Role of Ca^{++} in virus-induced by Sendai virus in chicken erythrocytes. J Cell Biol 78:465–479

Volsky DJ, Loyter A (1978a) An efficient method for reassembly of fusogenic Sendai virus envelopes after solubilization of intact virus with Triton X-100. FEBS Lett. 92:190–194

Volsky DJ, Cabantchik ZI, Beigel M, Loyter A (1979) Implantation of the isolated human erythrocyte anion channel into plasma membranes of Friend erythroleukemia cells by use of Sendai virus. Proc Natl Acad Sci USA 76:5440–5444

Wolosin JM (1980) A procedure for membrane protein reconstitution and the functional reconstitution of the anion transport system of the human erythrocyte membrane. Biochem J 189:35–44

Wolosin JM, Ginsburg H, Cabantchik ZI (1977) Functional characterization of anion transport system isolated from human erythrocyte membranes. J Biol Chem 252: 2419–2427

Modification

Chemical Modification of a Membrane Protein by Hydrophobic and Hydrophilic Labeling Reagents

H. SIGRIST and C. KEMPF

I. Introduction

The major transmembrane protein of the erythrocyte membrane – band 3 – has been chosen as an example of an integral membrane protein. This polypeptide is known to be involved in anion transport and has therefore been the focus of considerable research concerning both structure and function. Kinetic experiments and chemical modification studies have yielded valuable information on both of these aspects (Deutike 1977; Passow 1977; Cabantchik et al. 1978; Knauf 1979).

The hydrophobic probe phenylisothiocyanate (I) and its polar structural analog para-sulfophenylisothiocanate (II) are utilized for chemical modification of human erythrocyte band 3.

$$\text{(I)} \qquad\qquad\qquad\qquad \text{(II)}$$

The apolar reagent phenylisothiocyanate favorably partitions into the hydrophobic membrane phase ($\log P_{octanol/water} = 3.2$; Leo et al. 1971). The label may therefore interact with reactive nucleophiles present in this membrane domain. The polar $p\text{-}SO_3$-phenylisothiocyanate, however, is expected to preferentially label proteinaceous groups accessible from the aqueous phase. The opposing polar-apolar partition characteristics of the two reagents will be utilized to direct the probes to their respective favored domains for covalent interaction.

A. Reactivity of Arylisothiocyanates

Arylisothiocyanates form a covalent bond with nucleophilic groups only in their nonprotonated form ($RS^- \gg RO^- > RNH_2$) (Drobnika et al. 1977). For modification of membrane proteins the pH-controlled reactivity of proteinaceous nucleophilic groups can therefore be used for selective labeling by arylisothiocyanates (Sigrist and Zahler 1978; Sigrist et al. 1980).

Table 1. Interaction of arylisothiocyanates with amino acid side chain nucleophilic groups

Amino acid, nucleophilic group		pK_a	Reactive form of the nucleophile	Nucleophilic form present at neutral aqueous pH	
				Polar domain	Apolar domain
Cysteine	$-SH$	8.3	$R-S^-$	$R-SH/R-S^-$	$R-SH$
Serine, threonine	$-OH$	>14	$R-O^-$	$R-OH$	$R-OH$
Tyrosine	$-OH$	10.1	$R-O^-$	$R-OH$	$R-OH$
Lysine	$-NH_2$	10.5	$R-NH_2$	$R-NH_3^+$	$R-NH_2$

At neutral pH the amino group of lysine exposed to the aqueous phase is protonated and accordingly not reactive with either of the reagents. In contrast, the buried, bulk pH-independent amino functions may be in a reactive state (deprotonated), therefore making hydrophobic modification with phenylisothiocyanate feasible. Cysteine thiols exposed to the aqueous phase are partially deprotonated at neutral pH and, as a result, are reactive with both probes. The reaction product formed with thiols, however, is chemically distinguishable by its reversibility under alkaline conditions in the presence of excess reactive nucleophiles. Due to the extreme pK_a-value of tyrosine OH, the modification of this nucleophilic group is rather improbable in both the polar and apolar domain.

B.Aim of the Experiment

Under saturation conditions, 4 to 5 mol phenylisothiocyanate were found to be bound per mol band 3 (Sigrist et al. 1980). Binding of the polar probe p-sulfophenylisothiocyanate occurs in a protein fragment which topologically overlaps with the transmembrane 17,000 dalton peptide (Drikamer 1976, 1977). Labeling of whole erythrocytes with identical concentrations of pehnylisothiocyanate and p-sulfophenylisothiocyanate resulted in comparable inhibition of phosphate entry into the cells, although the anion-site directing group ($-SO_3^-$) is present only in the polar p-sulfophenylisothiocyanate (Kempf et al. 1979). It is the aim of the experiment to determine whether and to what extent p-sulfophenylisothiocyanate affects binding of phenylisothiocyanate to erythrocyte band 3 (differential labeling). Resulting information may thus contribute to both topological and functional considerations of the erythrocyte band 3 protein.

II. Equipment and Reagents

A. Equipment

ultracentrifuge (100,000 g), 10 centrifuge tubes (about 10 ml)
37°C water bath
magnetic stirrer
spectrophotometer
scintillation counter
glass tubes, plastic tubes, counting vials
Hamilton syringes, 25 μl, 50 μl, 100 μl
disposable micropipettes

B. Materials, Buffers, Reagents

human erythrocyte ghost membranes, prepared according to Dodge (Dodge
    et al. 1963), 4 mg protein/ml in 10 mM sodium phosphate buffer, pH 7.3
phenyl-[^{14}C]-isothiocyanate (Amersham) diluted with nonradioactive
    phenylisothiocyanate (Pierce, sequenal grade) to a specific activity of
    4 to 6 Ci/mol. Final concentration 0.1 M in ethanol
0.1 M p-sulfophenylisothiocyanate (Aldrich) in H$_2$O, freshly prepared solu-
    tion
10 mM sodium phosphate buffer, pH 7.3
10 mM sodium phosphate buffer, pH 7.3, containing 1% (w/v) sodium dode-
    cyl sulfate
solutions for protein determination in the presence of 0.1% sodium dodecyl-
    sulfate according to Lowry (Lowry et al. 1951)
solutions for 7.75% acrylamide-sodium dodecyl sulfate gel electrophoresis
    according to Fairbanks (Fairbanks et al. 1971), where 0.77% N,N'-diallyl-
    tartardiamide (Serva) is used instead of 0.29% N,N'-methylene-bisacryl-
    amide
0.088 M sodium periodate in H$_2$O
scintillation fluid: Triton X-100/Toluene/PPO/POPOP/acetic acid (1000:
    1000:5:0.2:40/v:v:w:w:v)

III. Experimental Procedures

In the first part of the experiment human erythrocyte ghost membranes will be modified by the arylisothiocyanates phenylisothiocyanate and p-sulfophenylisothiocyanate. Following sodium dodecyl sulfate gel electrophoresis of the labeled membranes, coomassie blue staining and destaining, the N,N'-diallyltartardiamide crosslinked acrylamide gels are ready for the second part of the experiment: the determination of the relative phenyl-[^{14}C]-isothiocyanate incorporation into electrophoretically separated band 3.

A. Differential Labeling of Erythrocyte Ghost Membranes

The following protein modification is performed in glass tubes. To avoid unspecific adsorption of the hydrophobic label glassware is used for both sampling and the reaction of phenylisothiocyanate with membranes.

Table 2. Protocol for differential labeling of erythrocyte ghost membranes

Tube No.	1	2	3	4	5	6	7	8	9	10
Erythrocyte ghosts, 4 mg/ml (ml)	0.5	0.5	0.5	0.5	0.5	0.5	0.5	0.5	0.5	0.5
p-Sulfophenyliso-thiocyanate, 0.1 M in H_2O (μl)	–	–	–	–	–	1	2.5	5	10	25
H_2O (μl)	25	25	25	25	25	24	22.5	20	15	0

Incubate for 30 min at 37°C, stirred suspensions. Phenyl-[^{14}C]-isothiocyanate or ethanol is added following the incubation

	1	2	3	4	5	6	7	8	9	10
Phenyl-[^{14}C]-isothio-cyanate, 0.1 M in ethanol (μl)	1	2.5	5	10	25	10	10	10	10	10
Ethanol (μl)	24	22.5	20	15	–	15	15	15	15	15

Within 5 min after the addition of phenyl-[^{14}C]-isothiocyanate duplicate samples (10 μl) are taken from each incubation mixture for the determination of total radioactivity (initial phenylisothiocyanate concentration). The tubes are covered and incubated (stirred suspensions) for 1 h at 37°C. The modified membrane are then transferred into centrifugation tubes. 10 mM sodium phosphate buffer, pH 7.3 is added (9 ml). The labeled membranes are then sedimented (20 min, 100,000 g, 4°C). The supernatant is discarded and the pelleted membranes are suspended with a syringe in 10 mM sodium phosphate buffer, pH 7.3. The described washing procedure is repeated twice. The final pellets are solubilized with 0.2 ml 10 mM sodium phosphate buffer, pH 7.3 containing 1% sodium dodecyl sulfate. For the determination of label incorporation into the membranes duplicate samples (10 μl) are withdrawn from the solubilized membrane for protein determination and radioactivity measurement. The initial concentration of phenylisothiocyanate used for membrane protein modification is calculated.

The phenylisothiocyanate incorporation into the washed modified membranes (cpm/mg membrane protein) may indicate the protective effect of p-sulfophenylisothiocyanate. Specific protein modification, however, can only be determined in the electrophoretically separated or chromatographically purified protein fraction.

B. Relative Phenyl-[^{14}C]-Isothiocyanate Incorporation into Band 3

Sodium dodecyl sulfate-polyacrylamide (7.75%) gel electrophoroesis is performed according to Fairbanks using 0.77% N,N'-diallyltartardiamide instead of 0.29% N,N'-methylene-bisacrylamide. This procedure allows a rapid solubilization of the acrylamide gel (Späth and Koblet 1979). Additionally, the relative protein content and radioactivity can be determined in the identical sample.

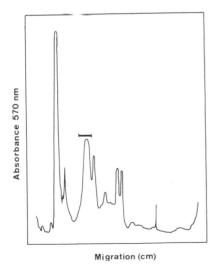

Fig. 1. Protein pattern of erythrocyte membranes separated on SDS-acrylamide gels described above. Position of band 3

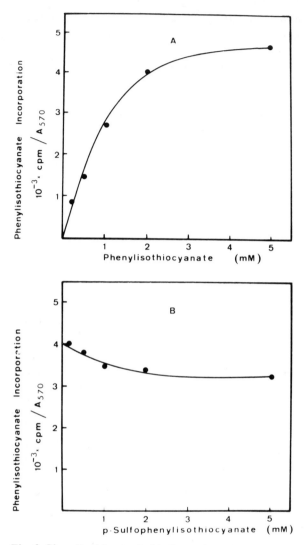

Fig. 2. Phenylisothiocyanate binding to band 3 protein and effect of p-sulfophenylisothiocyanate. Erythrocyte membranes were labeled with various phenylisothiocyanate
concentrations at pH 7.3 for 1 h, 37°C. The membranes were washed three times and
the label incorporation was determined in the electrophoretically separated band 3 by
analyzing the ratio of radioactivity to Coomassie blue staining intensity *(A)*. Prior to
labeling with 2 mM phenylisothiocyanate membranes were incubated with various
p-sulfophenylisothiocyanate concentrations *(B)*

In a duplicate analysis 100 μg solubilized arylisothiocyanate-labeled membrane protein is applied per gel. After electrophoresis the gels are Coomassie blue stained and destained. Duplicate samples of the stained Band 3 region of the gel (Fig. 1) are cut out (about 3 mm). The pieces are transferred into 3 ml plastic tubes and 0.8 ml 0.088 M sodium periodate is added. In about 30 min, solubilization of the gel takes place. Intermittent mixing with a vortex mixer is necessary. The optical density of the solubilized gel at 570 nm is determined (relative protein content). 2 ml scintillation fluid is then added to the total of the solubilized gel sample and, after mixing, assayed for radioactivity (1 min). The relative phenylisothiocyanate incorporation (cpm/A_{570}) into the electrophoretically separated band 3 is determined by the ratio of radioactivity to the corresponding optical density at 570 nm. Mean values of identically processed duplicate samples are then compiled in a graph as shown in Fig. 2.

VI. Comments

The effect of a p-sulfophenylisothiocyanate pretreatment on phenylisothiocyanate binding to membrane proteins is best described by the two extreme situations:

a) p-Sulfophenylisothiocyanate, at any concentration, does not affect phenylisothiocyanate binding. Such a result indicates that the phenylisothiocyanate binding sites are not accessible to the polar probe p-sulfophenylisothiocyanate. They are therefore most probably hydrophobically located. Such a result has been obtained by differential labeling of bacteriorhodopsin in purple membranes (Sigrist and Zahler 1980).

b) p-Sulfophenylisothiocyanate pretreatment induces a complete reduction of phenylisothiocyanate binding. Such a result strongly suggests the presence of interacting sites. The existence of common binding sites has to be certified by reversed experimental conditions: preincubation with phenylisothiocyanate followed by radio-labeled p-sulfophenylisothiocyanate.

Finally, various intermediate situations might result which are accordingly interpreted. The binding of phenylisothiocyanate to band 3, following a pretreatment of the erythrocyte ghost membranes with various p-sulfophenylisothiocyanate concentrations, results in a decrease of 20% to 30% phenylisothiocyanate binding (Fig. 2). In this case, the existence of a commonly affected site is concluded. Additionally, 70% to 80% of the sites accessible to phenylisothiocyanate are not affected by a pretreatment with 5 mM p-sulfophenylisothiocyanate. They are therefore most probably hydrophobically located.

References

Cabantchik ZI, Knauf P, Rothstein A (1978) The anion transport system of the red blood cell: The role of membrane protein evaluated by the use of 'probes'. Biochem Biophys Acta / Biomembrane Rev 515:239–302

Deutike B (1977) Properties and structural basis of simple diffusion pathways in the erythrocyte membrane. Rev Physiol Biochem Pharmacol 78:1–97

Dodge JT, Mitchell C, Hanahan D (1963) The preparation and chemical characteristics of hemoglobin free ghosts of human erythrocytes. Arch Biochem Biophys 110: 119–130

Drikamer KL (1976) Fragmentation of the 95,000-Dalton transmembrane polypeptide in human erythrocyte membranes. Arrangements of the fragments in the lipid bilayer. J Biol Chem 251:5115–5123

Drikamer KL (1977) Fragmentation of the band 3 polypeptide from human erythrocyte membranes. Identification of regions likely to interact with the lipid bilayer. J Biol Chem 252:6909–6917

Drobnika L, Kristian P, Augustin J (1977) The chemistry of the -NCS-group. In: Patai S (ed) The chemistry of cyanates and their thio derivatives, part 2. John Wiley and Sons, New York, pp 1002–1222

Fairbanks G, Steck TL, Wallach DFH (1971) Electrophoretic analysis of the major polypeptides of the human erythrocyte membrane. Biochemistry 10:2606–2616

Kempf Ch, Sigrist H, Zahler P (1979) Covalent modification of human erythrocyte band 3 phosphate transport inhibition by hydrophobic arylisothiocyanates. Experientia 35:937

Knauf P (1979) Erythrocyte anion exchange and the band 3 protein: Transport kinetics and molecular structure. In: Bronner F, Kleinzeller A (eds) Current topics in membrane transport, vol 12. Academic Press, London New York, pp 249–363

Leo A, Hansch L, Elkins D (1971) Partition coefficients and their uses. Chem Rev 71: 525–616

Lowry OH, Rosebrough NJ, Farr AL, Randall RJ (1951) Protein measurement with the folin phenol reagent. J Biol Chem 193:265–275

Passow H (1977) Anion transport across the red blood cell membrane and the protein in band 3. Acta Biol Med Ger 36:817–821

Sigrist H, Zahler P (1978) Characterization of phenylisothiocyanate as a hydrophobic membrane label. FEBS Lett 95:116–120

Sigrist H, Zahler P (1980) Heterobifunctional crosslinking of bacteriorhodopsin by azidophenylisothiocyanate. FEBS Lett 113:307–311

Sigrist H, Kempf Ch, Zahler P (1980) Interaction of phenylisothiocyanate with human erythrocyte band 3 protein. I. Covalent modification and inhibition of phosphate transport. Biochim Biophys Acta 597:137–144

Späth PJ, Koblet H (1979) Properties of SDS-polyacrylamide gels highly crosslinked with N,N'-diallyltartardiamide and the rapid isolation of macromolecules from the gel matrix. Anal Biochem 93:275–285

Crosslinking of an Amphiphilic Protein: Human Erythrocyte Membrane Acetylcholinesterase

U. BRODBECK and C.R. RÖMER-LÜTHI

I. Introduction

Crosslinking has become a method widely used to study the composition and symmetry of oligomeric proteins (Wold 1972; Peters and Richards 1977; Ji 1979; Das and Fox 1979). In this experiment an imidate and glutaraldehyde will be used as crosslinking reagents, both reacting with primary amino groups of proteins (Habeeb and Hiramoto 1968; Davies and Stark 1970). Imidates, under appropriate conditions are bifunctional reagents of well defined chain lengths (Davies and Stark 1970; Hajdu et al. 1976). Glutaraldehyde on the other hand forms multifunctional aggregates of undefined chain lengths (Richards and Knowles 1968; Monsan et al. 1975). As a consequence the number of crosslinks and thus the extent of crosslinking is expected to be smaller with imidates than with glutaraldehyde.

Human erythrocyte membrane acetylcholinesterase can be taken as general model of a membrane-bound, amphiphilic protein. Its high specific activity (Ott et al. 1975) which is not affected by the crosslinking reaction, even enhances the model character of this protein as the reaction products can easily be analyzed by a spectrophotometric enzyme assay.

Human erythrocyte membrane acetylcholinesterase is an integral membrane protein which is readily solubilized by detergents and then can be highly purified by affinity chromatography (Ott et al. 1975). On linear sucrose density gradients in presence of micellar detergent concentrations the purified enzyme shows one homogeneous form with a sedimentation coefficient of 6.5S. Due to hydrophobic interactions this form of the enzyme aggregates to multiple molecular forms up to 18S. Some of these aggregates have been characterized by analytical ultracentrifugation (Ott and Brodbeck 1978). The 6.5S form could be identified as dimeric species by crosslinking studies (Römer-Lüthi et al. 1979).

In the following experiments crosslinking (D) will be used to compare the aggregates and the 6.5S form of acetylcholinesterase. The molecular form(s) of this enzyme after reconstitution in large phosphatidylcholine vesicles are also examined by crosslinking.

The analysis of the resulting, crosslinked enzyme forms is performed by sucrose density gradient centrifugation in presence of micellar Triton X-100

(E) and by sodium dodecylsulfate polyacrylamide gel electrophoresis under reducing conditions (F).

The procedures outlined in this experiment, i.e., reconstitution, cross-linking, and methods of analysis of the crosslinked products should be applicable to other amphiphilic membrane proteins.

II. Equipment and Solutions

Recording photospectrometer with thermostat. Glass or plastic cuvettes of 3 ml. Quartz cuvettes of 1 ml. Unless otherwise indicated reagents are purchased by Fluka AG, CH-Buchs or by Merck, Darmstadt (FRG).

ad A: acetylcholinesterase assay mixture: solution containing 1 mM acetyl-thiocholine iodide, 0.125 mM 5,5'-dithio-bis-(nitrobenzoic acid), 0.05% Triton X-100 (Röhm and Haas, supplied by Bender and Hobein AG, CH-Zürich, Switzerland) in 100 mM phosphate buffer, pH 7.4

ad B: catalase assay mixture: solution containing 1 μl 30% H_2O_2 per ml of 100 mM phosphate buffer, pH 7.4

ad C: solution of 200 mM triethanolamine-HCl buffer, pH 8.5, containing 0.5% Triton X-100.
L-α-lecithin "ex egg" [Koch Light Laboratories, Colnbrook Bucks (GB)]
Amberlite XAD-2 [Serva, Heidelberg (FRG)] pretreated according to Holloway (1973). For experimental details see Brodbeck and Lüdi, this volume, p. 103)
nitrogen
system for gentle continuous mixing of a 2 ml sample without magnetic stirrer
glass filter and glass wool

ad D: solution of 200 mM triethanolamine-HCl buffer, pII 8.5 (containing 0.05% Triton X-100 for the purified enzyme in detergent)
10 mM Tris-HCl buffer, pH 7.0
dimethylsuberimidate (beware of hydrolysis!)
glutaraldehyde 25% in H_2O under N_2, purissimum for electron microscopy (Fluka AG, CH-Buchs) stabilized with glacial acetic acid (1 μl/ml) according to Monsan et al. (1975)
hydroxylamine

ad E: beef liver catalase [Boehringer, Mannheim (FRG)]

gradient mixer to allow the formation of linear sucrose density grad-
ients of 12 ml, containing 50 to 300 g sucrose/l in 20 mM Tris-
HCl buffer, 0.1 M NaCl, 0.1% Triton-X-100, pH 7.4
MSE 65 ultracentrifuge equipped with a 6 X 14 ml swing-out rotor

ad F: equipment for polyacrylamid gel electrophoresis
boiling water bath
10 mM Tris-HCl buffer, pH 7.0
acrylamide, N,N'-methylenebisacrylamide, Coomassie Brillant Blue
G-250 from Serva, Heidelberg (FRG)
high molecular weight calibration kit for gel electrophoresis (Pharmacia, Sweden)
sodium dodecylsulfate, mercaptoethanol

III. Experimental Procedures

A: Acetylcholinesterase activity is measured photospectrometrically according to Ellman et al. (1961).
Activity measurements are performed with a few microliters of enzyme
solution in 3 ml assay mixture and the increase of the absorbance is
followed at 412 nm at 25°C using a recording spectrophotometer.

B: Catalase activity is measured as described by Aebi (1974).
Activity measurements are performed with 1–2 μl of enzyme solution
in 1 ml assay mixture. The change in light transmission is followed at
240 nm and 25°C.

C: Reconstitution of purified human erythrocyte membrane acetylcholinesterase in egg phosphatidylcholine vesicles.
Phosphatidylcholine (8 mg) is dried with a stream of nitrogen and then
kept under vacuum for 1 h. The dried lipid is dissolved in 2 ml trietha-
nolamine buffer containing 0.5% Triton X-100 at 4°C. Amberlite
XAD-2 (0.6 g wet weight) is added and the mixture is gently agitated
for 1 min. Acetylcholinesterase (80 μg of protein) in 80 μl triethanol-
amine buffer containing 0.5% Triton X-100 is added. The solution is
gently mixed for 24 h at 4°C. Amberlite XAD-2 is then removed by
filtration through glass wool. The filtrate containing the reconstituted
enzyme is stored at 4°C under nitrogen.

D: Crosslinking of acetylcholinesterase.
The crosslinking reaction is performed at pH 8.5 in order to avoid
hydrolysis of the imidate and to stabilize the covalent bonds formed by

glutaraldehyde (Monsan et al. 1975). Both reagents are used in a final concentration of 20 mM. The enzyme solution is directly added to the dry imidate whereas glutaraldehyde, if necessary, is diluted in H_2O and then added to the enzyme solution. After 1 h at 25°C the crosslinking reaction is stopped by the addition of hydroxylamine to a final concentration of 200 mM. These samples are immediately used for sucrose density gradient centrifugation or have to be dialyzed against 10 mM Tris-HCl buffer, pH 7.0, at 4°C.

E: 100 μl samples of the crosslinked enzyme and 10 μl of catalase are layered on top of the linear sucrose density gradients and spun at 200,000 g, 4°C, for 15 h.

After centrifugation 500 μl fractions are collected from the bottom of the gradient and enzyme activities are measured in each fraction. Sedimentation values are calculated according to Martin and Ames (1961) based on the marker protein catalase (11.4S) as follows:

$$\frac{\text{sedimentation coefficient of sample}}{\text{sedimentation coefficient of marker protein}} = \frac{\text{distance of peak X moved from top of gradient}}{\text{distance of catalase peak moved from top of gradient}}$$

F: Sodium dodecylsulfate polyacrylamide gel electrophoresis is performed according to Fairbanks et al. (1971). Protein samples dialyzed against low ionic strength buffer are mixed with sodium dodecylsulfate and mercaptoethanol (10 g/l), sucrose (100 g/l) and a trace of bromophenol blue.

The protein samples are then incubated for 5 min in a boiling water bath and layered on top of the gels (about 10 μg protein per gel). Electrophoresis is carried out at 0.4 mA/mm^2 at 12°C for about 3 h. Coomassie blue-staining of the gels is performed as described in Fairbanks et al. (1971). Calibration of the gel system with a high molecular weight marker kit identifies the crosslinked oligomers of acetylcholinesterase.

IV. Results

A detailed characterization of the reconstituted system has been described in Römer-Lüthi et al. (1980). See also *Reconstitution of acetylcholine receptor from Torpedo marmorata,* U. Brodbeck and H. Lüdi, this volume.

The reconstituted vesicles may be visualized by freeze-fracture electron microscopy, as outlined in Römer-Lüthi et al. (1980). Gel filtration on a Sepharose 4B column shows a major peak of lipid phosphorus in the void volume of the column together with most of the acetylcholinesterase activity and some residual Triton X-100. Density gradient centrifugation of the vesicle preparation shows that both lipid and enzyme float on top of the gradient. A control experiment confirms that acetylcholinesterase, which is not lipid-associated, migrates toward the bottom of the gradient under the conditions employed (Römer-Lüthi et al. 1980).

Figure 1 shows the results of the sucrose density gradient experiment carried out with acetylcholinesterase crosslinked with 20 mM glutaralde-

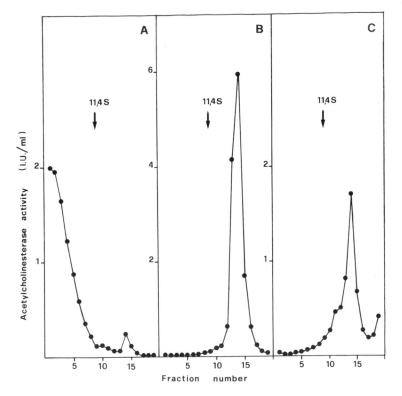

Fig. 1. Sucrose density gradient centrifugation of crosslinked acetylcholinesterase preparations. Crosslinking was performed with 20 mM glutaraldehyde. Analysis was performed on linear sucrose density gradients containing 5%–30% sucrose in 10 mM Tris-HCl, pH 7.4, 0.1 M NaCl and 0.1% Triton X-100. *A* Protein micelles of aggregated enzyme crosslinked in absence of Triton X-100; *B* 6.5S enzyme crosslinked in presence of 0.1% Triton X-100; *C* reconstituted enzyme crosslinked in absence of Triton X-100

hyde. It is seen that crosslinked protein micelles (aggregated enzyme) sediment toward the bottom of the gradient despite the presence of Triton X-100 (Fig. 1A). Thus the detergent is no longer able to dissociate the protein micelles due to intermolecular covalent bonds formed by glutaraldehyde. Figure 1B shows that the dimeric 6.5S enzyme is not crosslinked to higher aggregates when Triton X-100 is present in the reaction medium. The cross-linked reconstituted enzyme (Fig. 1C) that floats on top of the gradient when no detergent is present, migrates as 6.5S form in presence of Triton X-100.

Alternatively the crosslinked enzyme can be analyzed by sodium dodecyl sulfate gel electrophoresis (Fig. 2).

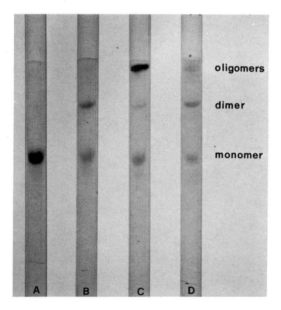

Fig. 2. Sodium dodecylsulfate polyacrylamide gel electrophoresis under reducing conditions of crosslinked acetylcholinesterase. *A* Control, uncrosslinked enzyme; *B* 6.5S enzyme crosslinked in presence of 0.1% Triton X-100; *C* protein micelles, crosslinked in absence of Triton X-100; *D* reconstituted enzyme crosslinked in absence of Triton X-100

IV. Comments

The methods of analysis presented in (E) and (F) give different information about the crosslinked enzyme. The sucrose density gradient centrifugation carried out in presence of Triton X-100 discriminates between the dimeric 6.5 S form of acetylcholinesterase (enzymatically active) and the crosslinked high molecular weight protein micelles of this dimer.

Electrophoresis under reducing conditions reveals the amount of uncross-linked monomer (80,000 molecular weight) and the enzyme forms with higher molecular weights, covalently interlinked by the crosslinking reagent(s).

Due to the multifunctional nature of glutaraldehyde an almost complete fixation of the aggregated enzyme forms can be achieved by this reagent (Figs. 1A and 2C) in contrast to the results obtained with suberimidate (Römer-Lüthi et al. 1979).

On the other hand no molecular forms larger than the dimer (6.5S) are found if the crosslinking reaction has been performed in presence of micellar Triton concentrations (Figs. 1B and 2B). This means that only intramolecular and no intermolecular crosslinking takes place under the conditions choosen.

A comparison of the crosslinked reconstituted enzyme with the Triton containing 6.5S form suggests that in both environments, detergent and lipids, acetylcholinesterase may exist mainly as dimeric 6.5S form (Figs. 1C and 2D).

Acknowledgment. Supported by grants no. 3.032-0.76 of the Swiss National Science Foundation.

References

Aebi H (1974) Bergmeyer HU (ed) Methoden der enzymatischen Analyse, 3rd edn, vol I. Verlag Chemie, Weinheim/Bergstr, pp 713–724
Das M, Fox CF (1979) Annu Rev Biophys Bioenerg 8:165–193
Davies GE, Stark GR (1970) Proc Natl Acad Sci USA 66:651–656
Ellman GL, Courtney DK, Andres V, Featherstone RM (1961) Biochem Pharmacol 7: 88–95
Fairbanks G, Steck TL, Wallach DFH (1971) Biochemistry 10:2606–2617
Habeeb AFSA, Hiramoto R (1968) Arch Biochem Biophys 126:16–26
Hajdu J, Bartha F, Friedrich P (1976) Eur J Biochem 68:373–383
Holloway PW (1973) Anal Biochem 53:304–308
Ji H (1979) Biochem Biophys Acta 559:39–69
Martin RG, Ames BN (1961) J Biol Chem 236:1372–1379
Monsan P, Puzo G, Mazarguil H (1975) Biochimie 57:1281–1292
Ott P, Brodbeck U (1978) Eur J Biochem 88:119–125
Ott P, Jenny B, Brodbeck U (1975) Eur J Biochem 57:469–480
Peters K, Richards F (1977) Annu Rev Biochem 46:524–551
Richards FM, Knowles JR (1968) J Mol Biol 37:231–233
Römer-Lüthi CR, Hajdu J, Brodbeck U (1979) Hoppe-Seyler's Z Physiol Chem 360: 929–934
Römer-Lüthi CR, Ott P, Brodbeck U (1980) Biochem Biophys Acta 601:123–133
Wold F (1972) Methods Enzymol 25:623–651

Labeling of Membranes from Within the Lipid Core: Use of a Photoactivatable Carbene Generator

J. BRUNNER

I. Introduction and Aims

Labeling of membranes from within the lipid core aims at identifying those polypeptide segments of membrane proteins that are buried in the lipid bilayer. This method, therefore, represents a useful complement to the more established surface labeling techniques of membranes.

Since most of the polypeptide segments that penetrate the lipid core are nonpolar and chemically inert, their modification requires a highly reactive reagent. Such considerations have led to the development of compounds that are chemically unreactive in the dark, but generate highly reactive intermediates when exposed to ultraviolet irradiation. In principle, some intermediates thus generated are capable of reacting with the full range of amino acid side chains including aliphatic residues. Labeling of membranes can be restricted to intrinsic proteins if a reagent is used that partitions highly in favor of the lipid core of the membrane (Brunner 1981).

Using photoactivatable probes for membrane structure has inherent problems:

1. The photoactivation of the reagent must occur under conditions that do not cause photooxidation or other damage to the membrane.
2. The reagent should react "instantaneously", which means that the reactive intermediate must not rearrange to less reactive and possibly more polar intermediates that can diffuse out of the bilayer and react from the aqueous environment.
3. Specific binding of the reagent or steric exclusion must not occur.
4. The photogenerated intermediate should be of very high reactivity.

The principal goal of this article is to describe some of the methodological aspects of hydrophobic membrane labeling. This includes a description of the preparation of a reagent of high specific radioactivity and the procedures of sample preparation and photolysis.

II. Equipment and Solutions

laboratory facilities for handling ^{125}iodine in quantities of 1 mCi (C-type
  laboratory)
photolysis apparatus
thin-layer chromatography equipment [Kieselgel 60 F_{254} plates from Merck,
  ultraviolet lamp (254 nm)]
reactivials (conical) with "Mininert" valves from Pierce
Hamilton syringes
water baths ($0°C$, $50°C$)
centrifuges (Sorval, MSE)

Reagents and solutions:

3-trifluoromethyl-3-(m-formylaminophenyl)diazirine (TFD), 0.5 M in methanol (Brunner et al. 1981)
hydrochloric acid 10 M (HCl)
sulfuric acid 3 M (H_2SO_4)
sodium nitrite 1 M ($NaNO_2$)
sodium iodide (NaI)
sodium hydrogen sulfite ($NaHSO_3$) 5%
carbontetrachloride (CCl_4)
n-hexane
ethanol
sodium ^{125}iodide (carrier-free) from EIR, Würenlingen (delivered as a solution of approximately 1 mCi/2 μl

III. Experimental Procedures

A. Preparation of [^{125}I]3-Trifluoromethyl-3-(m-iodophenyl)diazirine (TID)

All operations involving diazirines must be performed in the absence of sun-
light; the diazirines are not sensitive, however, to fluorescent room illumination. Since the radioactive TID slowly decomposes, the radioiodine is introduced into a suitable precursor immediately (or a few days) before the
labeling experiments are performed. Figure 1 shows the synthetic pathway
(Brunner et al. 1981).

Reaction 1. In a test tube, TFD (200 μl, 0.5 M) in methanol is mixed with
HCl (100 μl, 10 M) and the solution kept at $25°C$ for 30 min. Methanol and
HCl are then evaporated by means of a stream of dry nitrogen and the
residue dissolved in 800 μl of 3 M H_2SO_4.

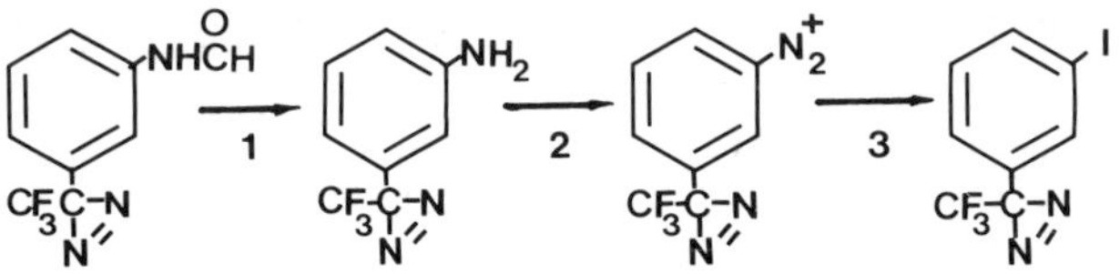

Fig. 1. Reaction scheme for the preparation of [^{125}I]3-trifluoromethyl-3-(m-iodophenyl) diazirine

Reaction 2. The arylamine is diazotized at 0°C with 5 to 10 μl portions of NaNO$_2$ (1 M) until a lasting (> 30 min) positive reaction is observed on KI-starch indicator strips.

Reaction 3. In a Reactivial, 1 mCi of Na[^{125}I] (carrier-free) is diluted with NaI so that 15 μl of a Na [^{125}I]solution is obtained that has a specific radioactivity of 2 Ci/mmol. The vial is placed on ice and 15 μl of the diazonium salt solution (Reaction 2) is injected through the silicon rubber septum of a Mininert valve with a precooled Hamilton syringe. The sealed vial is transferred into a water bath (50°C) and kept for 30 min. After cooling to room temperature, 50 μl of a 5% solution of NaHSO$_3$ is added (SO$_2$ gas evolved is released through the needle of the syringe) and the turbid solution is kept at room temperature for an additional 10 min. Then, 5 μl of CCl$_4$ is added and the valve replaced by a screw-cap equiped with a Teflon gasket. Following gentle vortexing, the organic phase which contains the reagent is collected in the tip of the conical vial by centrifugation (MSE table centrifuge; 5 min, 1000 rpm). The organic solution is transferred onto a TLC plate (2 × 15 cm) by means of a Hamilton syringe and the chromatogram developed in hexane. The band containing the radioactive reagent can be precisely located on the TLC plate from its fluorescent quenching when viewed under a UV lamp. This band of silica gel is scraped off and the [^{125}I] TID extracted from the silica gel with approximately 300 μl of ethanol ([^{125}I] TID Stock solution; 2–3 μCi/μl).

Precaution. TID is slightly volatile; the purification of the reagent must be performed in a glove-box.

B. Preparation of Photolysis Samples

In a glass test tube (with glass stopper), 1.0–1.2 ml of human erythrocyte membranes (1–2 mg protein/ml) is thoroughly deoxygenated by bubbling through the membrane dispersion a gentle stream of nitrogen (30 min).

An appropriate volume of the ethanolic solution of [^{125}I] TID (0.5 μCi is sufficient for the determination of the time course of the photochemical labeling) is injected with a syringe and the glass tube sealed. The incubation mixture is equilibrated at 0°C for 15–30 min and then transferred into the sample cell of the photolysis apparatus. The cuvette is briefly flushed with nitrogen and sealed (see also Comments).

C. Time Course of the Incorporation of Label into Red Cell Membranes

The apparatus used for irradiation is schematically shown in Fig. 2. The membrane sample is photolyzed for successive periods of time (15 s). At each interval, 50 μl aliquots of the photolyzed membranes are removed and diluted with 1.95 ml of sodium phosphate buffer (5 mM; pH 8) containing 1% bovine serum albumin. The membranes are sedimented at 15,000 g for 30 min in glass centrifuge tubes. Supernatants are removed carefully and this washing of the membranes is repeated in a similar manner for at least three more times and then twice with albumin-free phosphate buffer. The washed cells containing the covalently bound label are transferred into disposable polystyrene tubes and the radioactivity is counted.

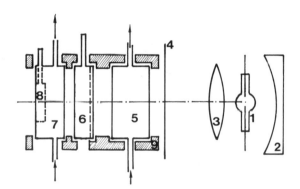

Fig. 2. Scheme of photolysis apparatus: *1* Light source (medium or high pressure mercury lamp); *2* reflector; *3* lense; *4* shutter; *5* infra-red filter (circulating cold water); *6* filter combination (saturated copper sulfate solution, 20 mm); *7* thermostated cell; *8* sample cuvette; *9* tube

IV. Comments

From the experiment described, the time period required for complete photolysis of the diazirine under the experimental conditions of irradiation can be derived. For photolabeling studies, the photolysis time should be sufficient to cause complete decay of the diazirine. Prolonged irradiation, however, results in photo-induced crosslinking of membrane components.

The time course of the photolabeling process critically depends on the light source used (e.g., medium or high pressure mercury lamp, or xenon lamp) and the filter combination, between the light source and the sample cuvette, which is required to screen out the short wavelength ultraviolet irradiation.

It is important to realize that handling of radioiodinated compounds requires adequate precautions. This is important in particular when the labeled compound is slightly volatile as is TID. Flushing of solutions containing the reagent with nitrogen must be strictly avoided unless the nitrogen can subsequently be decontaminated by passing through an adsorbent. Since TID is a very hydrophobic compound it can be adsorbed efficiently by almost any forms of plastic and rubber (for this reason, parafilm should not be used for sealing).

Acknowledgments. The author wants to thank SNF, Berne, for financial support and Prof. G. Semenza for fruitful discussions.

References

Brunner J (1981) Labelling of the hydrophobic core of membranes. Trends in Biochem Sci 6:44–46

Brunner J, Semenza G. Selective labeling of the hydrophobic core of membranes with [^{125}I]3-trifluoromethyl-3-(m-iodophenyl)diazirine, a carbene generating reagent. To be submitted

Recommended Reading

Bayley H, Knowles JR (1977) Photoaffinity labeling. Methods Enzymol 46:69–144

Bercovici T, Gitler C (1978) 5-[^{125}I]Iodonaphthyl-1-azide, a reagent to determine the penetration of proteins in the lipid bilayer of biological membranes. Biochemistry 17:1483–1489

Chowdhry V, Westheimer FH (1979) Photoaffinity labeling of biological systems. Annu Rev Biochem 48:293–325

Goldman DW, Pober JS, White J, Bayley H (1979) Selective labelling of the hydrophobic segments of intrinsic membrane proteins with a lipophilic photogenerated carbene. Nature (London) 280:841–843

Characterization – Spectral Techniques

Applications of Spin-Labels to Biological Systems

C. BROGER and A. AZZI

I. Introduction and Aims

Labeling techniques have been used for many years in the study of biological systems. The labels can be radioactive isotopes, fluorescent or colored molecules. Those used in electron spin resonance spectroscopy are free radicals (spin-labels). Their ESR spectrum is sensitive to the distance between labels (concentration effect), to the polarity of the environment, to the type of motion (isotropic, anisotropic) and to the velocity of motion.

We intend to illustrate in this chapter some of the applications of spin-labeling in biochemistry by simple and basic experiments.

Most spin-labels used nowadays contain the unpaired electron in a nitroxide group of the general formula shown in Fig. 1. The methyl groups make the free radical extremely stable. R_1 and R_2 are side groups which provide the nitroxide with specific reactivity and/or physicochemical properties (see Figs. 2, 3, and 4). All these compounds are commercially available.

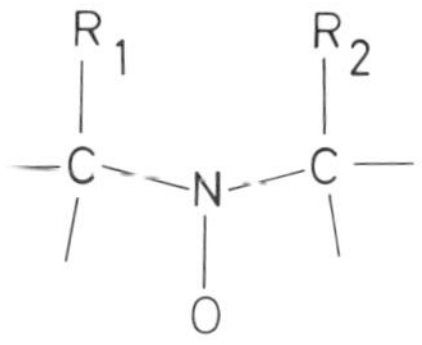

Fig. 1. General formula of nitroxide spin-labels

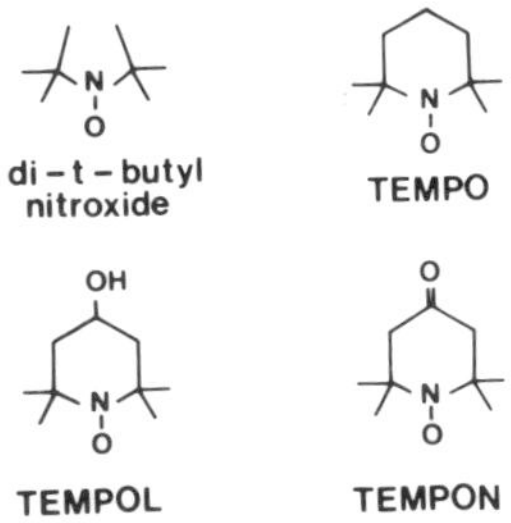

Fig. 2. Spin-probe: used to probe environment (especially polarity)

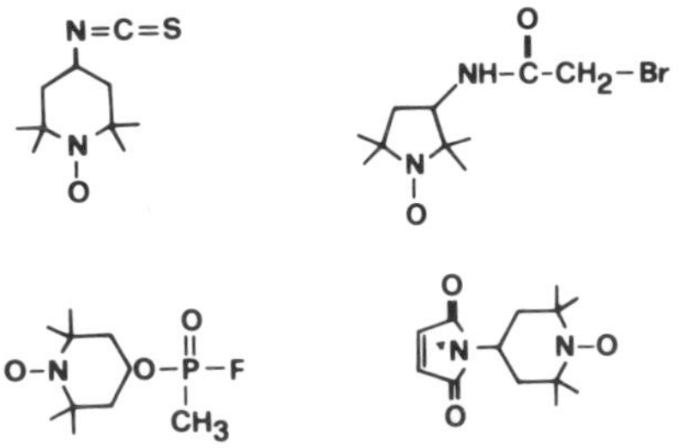

Fig. 3. Spin-labels containing reactive groups for -SH, -NH$_2$, -OH and other residues, for example of amino acids

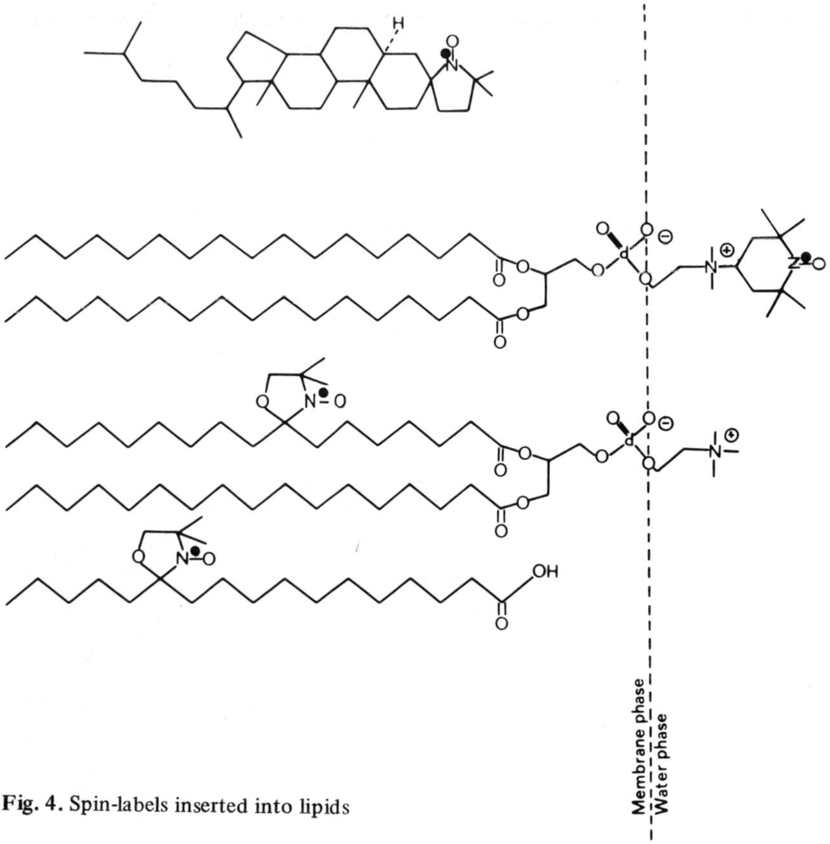

Fig. 4. Spin-labels inserted into lipids

II. Theory

References: Berliner (1976, 1979); Wertz and Bolton (1972); Jost and Griffith (1978)

A. Classical ESR Spectroscopy

Quantum mechanics show that as a free electron in an applied magnetic field has a spin of 1/2 it can assume only two spin states with components of the spin angular momentum along the direction of the applied magnetic field having the quantum numbers

$$M_S = \pm 1/2$$

The charge of the electron gives rise to a magnetic moment which has components along the direction of the applied magnetic field of

$$\mu_z = -g\beta M_S$$
(negative sign because of the negative charge of the electron)

where g is the so-called g-value (2.00232 for a free electron, around 2 for nitroxide spin-labels, possibly different from 2 for transition metal ions), β is the Bohr magneton ($9.274 \cdot 10^{-21}$ erg/Gauss).
Substituting $\pm 1/2$ for M_S yields

$$\mu_z = -1/2\, g\, \beta \quad \text{and} \quad \mu_z = 1/2\, g\, \beta \text{ for the two states}$$

The energy of a dipole in a magnetic field is defined as

$$E = -\mu H \quad \text{where H is the magnetic field strength}$$

The energies of the electron in the two states (the Zeemann energies) are thus

$$E_1 = +1/2\, g\, \beta\, H \quad \text{and} \quad E_2 = -1/2\, g\, \beta\, H$$

The energy difference between these two levels is

$$\Delta E = g\, \beta\, H$$

Since g and β are constants, the separation between the Zeeman levels increases linearly with the magnetic field strength. Transitions between the levels can be induced by an electromagnetic field of the appropriate frequency ν (in the microwave range) so that the photon energy $h\nu$ matches the energy level separation ΔE.

Then $\Delta E = h\nu = g\, \beta\, H_r$

where h is the Planck constant ($6.626 \cdot 10^{-27}$ erg s) and H_r is the magnetic field strength at which the resonance condition is met. Typical values for ν and H_r are 9.4 GHz and 3300 Gauss, respectively.
In principle, an ESR spectrum could be scanned by sweeping either the magnetic field H or the microwave frequency ν. For practical reasons it is much easier to sweep the magnetic field H, keeping the microwave frequency constant.
The coupling of the electron spin angular momentum with the orbital angular momentum gives rise to the tensor g, which is slightly dependent on

the polarity of the solvent. In oriented systems (e.g., crystals) its value is different for different orientations of the crystal with respect to the direction of the applied magnetic field (anisotropy). The molecular coordinate system for nitroxide spin-labels is shown in Fig. 5. The value of g measured with the applied magnetic field along the x-axis of the molecular coordinate system is called g_{xx} • g_{yy} and g_{zz} are defined in a similar way.

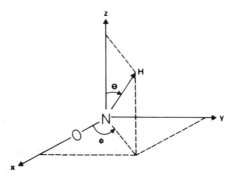

Fig. 5. Molecular coordinate system of nitroxide spin-labels. The *x-axis* lies along the N-O bond. The *z-axis* lies along the p_z orbital of nitrogen. The *y-axis* is perpendicular to the x- and z-axes. *H* is the direction of the applied magnetic field. θ is the angle between the magnetic field and the z-axis. ϕ is the angle between the x-axis and the projection of H in the x-y-plane

If the magnetic field is not applied along one of the main axes, g can be calculated for a certain orientation from the three main values of g as follows:

$$g = g_{xx} \sin^2 \theta \cos^2 \phi + g_{yy} \sin^2 \theta \sin^2 \phi + g_{zz} \cos^2 \phi$$

In an isotropic system with fast molecular motion the g value measured is the average of the g values along the main axes:

$$g_0 = 1/3 \, (g_{xx} + g_{yy} + g_{zz})$$

In nitroxide spin-labels the unpaired electron is located near the nitrogen atom. The most abundant isotope of nitrogen [^{14}N] possesses a nuclear spin of I = 1. Therefore the components of the spin along the direction of the applied magnetic field can assume the quantum numbers M_I = + 1, 0 and − 1. The magnetic moment associated with the nuclear spin produces, at the site of the electron, a small additional magnetic field either in the direction of the applied magnetic field or opposite to it. In the case of M_I = 0 no additional field is produced. Therefore the resonance condition for the electron is met at three different values of the applied magnetic field, depending on the state of the adjacent nitrogen nucleus (hyperfine splitting). Since the probability of finding the nucleus in one of the three states is one third, all three absorption lines have the same height (see Fig. 6).

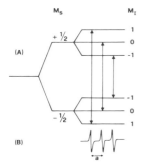

Fig. 6. A Energy levels and transitions for the unpaired electron in the nitroxide molecule. **B** First derivative spectrum. *a* is the hyperfine splitting constant

In normal ESR spectroscopy not the absorption lines are displayed but their first derivative. This is the case because a small amplitude 100 kHz modulation field is superimposed on the applied magnetic field. Detection also occurs at 100 kHz. As can be seen from Fig. 7 the resulting detector current is proportional to the slope of the absorption curve. A modulation frequency of 100 kHz and the in-phase detection are mainly used to improve the signal-to-noise ratio.

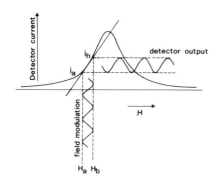

Fig. 7. Effect of the small-amplitude 100 kHz field modulation on the detector current. The applied magnetic field is modulated between the limits H_a and H_b. The corresponding detector current varies between i_a and i_b

The hyperfine splitting constant can be measured from the spectrum as the distance between two peaks (in Gauss). Hyperfine splitting arises from dipole-dipole interaction between electron and nucleus and is also influenced by the polarity of the solvent. Its value decreases with decreasing polarity of the solvent. The spin-label TEMPO, e.g., if added to a lipid suspension in water, partitions between the aqueous and lipid phase. The resulting spectrum is the superimposition of the spectra of TEMPO in water and that in the lipid (Fig. 8). The high field peak is split into two lines. A ratio r can be defined:

$$r = \frac{H}{H + P}$$

where H is the height of the peak from TEMPO in the hydrophobic phase and P is the height of the peak from TEMPO in the polar phase. r is called TEMPO spectral parameter.

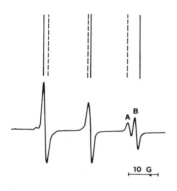

Fig. 8. Partition of TEMPO between a polar and a hydrophobic phase. *Top: Solid lines* spectrum of TEMPO in water. *Dotted lines* spectrum of TEMPO in lipid. *Bottom:* Superimposition of the two spectra. Due to the differences of the g- and a-values in the two phases only the high field peak is resolved into two components. (*A* = hydrophobic; *B* polar phase)

As for g, in ordered systems the value which is measured for a depends on the orientation of the system with respect to the applied magnetic field. The three values measured with the magnetic field along the molecular main axes are called a_{xx}, a_{yy}, and a_{zz}. a_{zz} is always much bigger than a_{xx} and a_{yy}, which have about the same value.

The value of a along an intermediate axis can be calculated as follows:

$$a = [(a_{xx})^2 \sin^2 \theta + (a_{zz})^2 \cos^2 \theta]^{1/2}$$

assuming $a_{xx} \sim a_{yy}$.

In the case of rapid random motion the anisotropic values are averaged yielding:

$$a_0 = 1/3 (a_{xx} + a_{yy} + a_{zz})$$

The resulting spectrum consists of three sharp lines.

The other extreme situation would be a frozen sample with random arrangement of the spin-labels. In this case the sepctrum is the superimposition of spectra of molecules randomly oriented with respect to the applied magnetic field (rigid glass spectrum). The distance between the two extreme lines corresponds to 2 a_{zz} (Fig. 9).

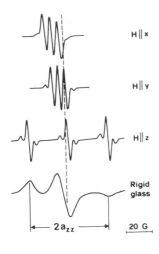

Fig. 9. Spectra of a crystal of spin-label measured with the applied magnetic field in the direction of the three main molecular axes and of a frozen sample containing spin-labels randomly arranged

The shape of the spectrum of isotropically moving spin-labels depends on the rotational correlation time of the molecules between about 10^{-7} s (the so-called rigid glass or immobilized spectrum) and 10^{-11} s (three sharp lines of equal height, mobile spectrum). Intermediate states are shown in Fig. 10. They can be reached by changing the viscosity and/or temperature of the system.

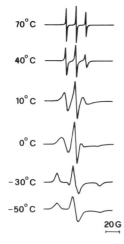

Fig. 10. Spectra of spin-labels rotating isotropically. The rotational correlation time is increasing from top to bottom from about 10^{-11} to 10^{-7} s

Approximate rotational correlation times can be calculated from these spectra by the following formula (Stone et al. 1965):

$$\tau_c = 6.5 \cdot 10^{-10} \cdot \Delta H_0 \left(\sqrt{\frac{h_0}{h_{-1}}} - 1 \right) \text{ sec}$$

where ΔH_0 is the linewidth of the central line in Gauss (distance between maximum and minimum deflection of the recorder) and h_0 and h_{-1} are the intensities of the central and high field lines. The formula holds for rotational correlation times between about 10^{-11} and 10^{-9} s. Spectra of spin-labels with isotropic motion occurring in times longer than 10^{-9} s can be described in terms of the overall splitting which increases with immobilization.

Molecules containing a main axis around which rotation occurs preferentially, e.g., fatty acids, are a special case. In a lipid bilayer these molecules undergo anisotropic motion (fast rotation around the long axis which coincides with the z-axis of the spin-label and almost no rotation around axes perpendicular to the long axis). In the case of such axial symmetry, maximal hyperfine splitting occurs when the applied magnetic field is parallel to the fatty acid chain and minimal splitting when it is perpendicular to the chain. The subscripts of g and a can be adapted to this situation in the following way: g_{zz} and a_{zz} are changed into $g_{\parallel}$ and $a_{\parallel}$; while g_{xx}, g_{yy} and a_{xx}, a_{yy} are changed into $g_{\perp}$ and $a_{\perp}$. If the rotation occurred exclusively about the long axis of the fatty acid, $a_{\parallel}$ would be equal to a_{zz} and $a_{\perp}$ equal to a_{xx} and a_{yy}. Since there is always a smaller or bigger part of rotation around the axes perpendicular to the fatty acid chain, $a_{\parallel}$ will be somewhat smaller than a_{zz} and $a_{\perp}$ will be bigger than a_{xx} and a_{yy}. The shape of the spectrum of a fatty acid rotating preferentially around the long axis is shown in Fig. 11.

The values of $a_{\parallel}$ and $a_{\perp}$ can be measured from the spectrum. By comparison with a_{xx}, a_{yy} and a_{zz} one can infer to what extent the motion of the fatty acid molecules occurs only in the long axis and consequently how much of the lipid bilayer is ordered. A measure of such ordering is the

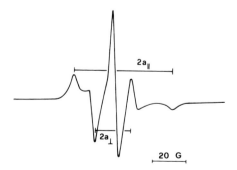

Fig. 11. Spectrum of an anisotropically rotating fatty acid in a lipid bilayer

"order parameter" defined as follows:

$$S = \frac{a_\parallel - a_\perp}{a_{zz} - 1/2 \, (a_{xx} + a_{yy})} \cdot \frac{1/3 \, (a_{xx} + a_{yy} + a_{zz})}{1/3 \, (a_\parallel + 2a_\perp)}$$

The factor containing the ratio between the isotropic hyperfine splitting constants is included to correct for different polarities in a crystal (where a_{xx}, a_{yy} and a_{zz} are measured) and the bilayer.

S = 0 means that the motion of the spin-label is isotropic
S = 1 means that there is only rotation around the long axis of the fatty acid

If the distance between two spin-labels is decreased, a dipolar interaction between them will occur. By further diminishing the distance, the wave functions of the electrons will overlap and spin-spin exchange will occur. The spectrum will then consist of only one broad peak (see Fig. 12).

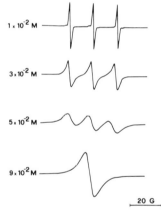

1×10^{-2} M

3×10^{-2} M

5×10^{-2} M

9×10^{-2} M

20 G

Fig. 12. Dependence of the ESR spectrum on the concentration of spin-label. The concentration is increasing from top to bottom

B. Saturation Transfer ESR Spectroscopy

As mentioned above, with classical ESR spectroscopy rotational correlation times between 10^{-11} and 10^{-7} s can be measured. It can easily be calculated from the Debye equation for Brownian rotational diffusion

$$\tau_c = \frac{4 \, \pi \, \eta \, r^3}{3 \, kT}$$

[where η is the viscosity of the sample (in Poise), r is the radius of the rotating molecule, k is the Boltzmann constant, T is the absolute temperature] that macromolecules, e.g., proteins, rotate with correlation times in the order of magnitude of micro- and milliseconds.

In the last years saturation transfer ESR spectroscopy was developed to study such so-called "very slow motions".

The principle of saturation transfer is the following: As shown above, different absorption peak positions in the spectrum correspond to different orientations of the spin-labels with respect to the applied magnetic field. If the sample is irradiated with a microwave field, stronger than used for normal ESR spectroscopy, the spin-labels in the proper orientation are partially saturated. While still in this saturated state they rotate into different orientations and consequently into different positions of the spectrum. In this way saturation is diffusing from one position in the spectrum into others. If the rotational correlation time is comparable to the time in which saturation is lost by the spin-system, the spin-lattice relaxation time ($T_1 \sim 10^{-5}$ s for nitroxides), the shape of the spectrum will be strongly dependent on the rotational correlation time. Practically, with this method, rotational correlation times can be measured between 10^{-7} and 10^{-3} s.

There are several methods of observing saturation transfer, e.g., electron-electron double resonance or absorption spectroscopy, using either first or second harmonic detection both in- or out-of-phase, or dispersion first harmonic detection in- or out-of-phase.

Second harmonic out-of-phase absorption spectra are easy to measure and simultaneously show a rather large dependence on the rotational correlation time. This technique is also employed in the present experiment.

Modulation of the applied field is in this case used not only to improve the signal-to-noise ratio, but is, because its period is of the same order of magnitude as T_1 and τ_c, together with the rotational diffusion, very important to bring the electrons through the resonance condition. Phase-sensitive detection (90° shifted with respect to the modulation) is used to select the portion of the signal that arises from saturation effects.

Calibration of the rotational correlation times is made using standard solutions of spin-labeled hemoglobin having different viscosities.

III. Equipment and Chemicals

A. Equipment

ESR spectrometer Varian E-104A
variable temperature controller
flat quartz cell (200 μl)
capillaries made from Pasteur pipettes

Settings of the ESR spectrometer:
a) normal experiments: modulation frequency 100 kHz
                           modulation amplitude 1 Gauss
                           microwave power 5–10 milliwatts
b) saturation transfer: second harmonic mode
                           (modulation frequency 50 kHz
                           detection frequency 100 kHz)
                           90° out-of-phase detection
                           modulation amplitude 5 Gauss
                           microwave power 63 milliwatts

At microwave powers well below saturation (about 1 milliwatt) the phase is
adjusted so that the out-of-phase signal on the recorder is minimal.

B. Chemicals

TEMPO 1 mg/ml in ethanol
TEMPON 25 mg/ml in water
spin-label stearyl acids containing the spin-label in the positions C-5, C-12,
    and C-16 2 mg/ml in ethanol
4-maleimide-2,2,6,6-tetramethylpiperidinoxyl (spin-label maleimide)
dimyristoyl phosphatidylcholine 100 mg/ml in chloroform
dipalmitoyl phosphatidylcholine 100 mg/ml in chloroform
hemoglobin 6 mM in 10 mM Na phosphate buffer pH 6.7
cytochrome c oxidase from beef heart 15 mg/ml in 100 mM Na phosphate
buffer pH 7.0, containing 0.5% Tween
100 mM Na phosphate buffer pH 7.0
100 mM Na phosphate buffer pH 7.0, containing 0.5% Tween
glycerol
cyclohexane
Sephadex G-25 coarse in chromatography columns (10 × 1 cm)

IV. Experiments

A. Polarity Effects

a) put 20 μl of TEMPO solution into two tubes
evaporate the solvent under nitrogen
add 100 μl of water to one of the tubes and 100 μl of cyclohexane to the other

put the samples into capillaries
measure the spectra of the samples in the ESR spectrometer
read a_0 from both spectra

b) Measurement of the melting point of a dimyristoyl phosphatidylcholine suspension in water (Shimshick and McConnell 1973)

The partition of TEMPO between the lipid and the water phase in a lipid suspension in water is temperature-dependent. Below the melting point of the lipid the spin-probe is excluded from the lipid due to its crystalline structure. When the lipid melts, TEMPO becomes more soluble in the lipid. The solubility is measured by means of the TEMPO spectral parameter at different temperatures.

put 100 μl of dimyristoyl PC solution into a tube
add 4 μl of TEMPO solution
evaporate the solvents
add 100 μl of water and shake vigorously at about 30–40°C (above the transition temperature of the lipid) until the suspension is homogeneous
put the sample into a capillary and measure it at different temperatures between 0 and 30°C
calculate the TEMPO spectral parameter for all the temperatures

$$r = \frac{H}{H + P} \qquad \begin{array}{l} H = \text{spin-probe in hydrophobic phase} \\ P = \text{spin-probe in polar phase} \end{array}$$

plot the spectral parameter against the temperature. The point with the biggest slope is the melting point of the lipid. Possibly you can also see a phase transition of the lipid before its real melting point.

B. Concentration Effects

Prepare the following mixtures in four tubes:

a) 100 μl TEMPON solution + 250 μl water
b)   50 μl TEMPON solution + 100 μl water
c)   50 μl TEMPON solution +   50 μl water
d)   50 μl TEMPON solution

put 50 μl from each tube into a capillary and scan the ESR spectrum
calculate in each case an average distance between two TEMPON molecules (assuming each one is confined to a cube)

At what intermolecular distances does the spectrum start to show dipolar interactions between spins and spin-spin exchange?

Example of an application (Sigrist-Nelson and Azzi 1979)

The F_0 proteolipid of ATPase was labeled with the spin-label analog of dicyclohexylcarbodiimide, NCCD. As long as the protein is held in solution it shows a normal ESR spectrum. Upon reconstitution of the proton channel, the ESR spectrum shows spin-spin exchange, indicating close distance between several polypeptides: The proton channel is formed by more than one subunit.

C. Isotropic Motion

put 50 μl of TEMPO solution into one tube
evaporate the solvent
add 100 μl of glycerol
heat with hot water and shake to dissolve the spin-probe homogeneously
put the solution into a capillary and measure the ESR spectrum at several
   temperatures between $-50°$ and $+70°C$
calculate, where possible, the rotational correlation times using the formula

$$\tau_c = 6.5 \cdot 10^{-10} \cdot \Delta H_0 \left(\sqrt{\frac{h_0}{h_{-1}}} - 1 \right) s$$

where ΔH_0 is the line width of the central line in Gauss and h_0 and h_{-1} are the intensities of the central and the high field lines.

Example of an application (Leute et al. 1972)

Spin-labeled morphine and antibodies against morphine are prepared. The antibodies are mixed with a proper amount of spin-label morphine so that the antibodies are saturated. Because of binding to the protein, the motion of the morphine is largely diminished and the ESR spectrum shows a rotational correlation time of 10^{-7} s or longer. Then, for example, urine is added. If it contains morphine the spin-labeled morphine is displaced from the antibodies and free spin-label with a much shorter rotational correlation time appears in the ESR spectrum. The intensity of its signal is proportional to the morphine concentration in the urine.

D. Anisotropic Motion

put 20 μl of dipalmitoyl PC into 3 tubes
to each tube add 3 μl of one of the spin-labeled fatty acids
evaporate the solvents

add 200 μl of water

heat with warm water and shake

measure the samples using the flat quartz cell

calculate the order parameter for each of the fatty acids according to the
    formula

$$S = \frac{a_{\parallel} - a_{\perp}}{a_{zz} - 1/2\,(a_{xx} + a_{yy})} \cdot \frac{1/3\,(a_{xx} + a_{yy} + a_{zz})}{1/3\,(a_{\parallel} + 2a_{\perp})}$$

use the following hyperfine splitting constants:

$a_{xx} = 5.9$ Gauss

$a_{yy} = 5.4$ Gauss

$a_{zz} = 32.9$ Gauss

plot the order parameter against the number of the carbon atom carrying
the label.

E. Saturation Transfer Spectroscopy (Thomas et al. 1976; Hyde 1978)

1. Preparation of Spin-Labeled Hemoglobin (McCalley et al. 1972)

Three mg of solid spin-label maleimide (twofold excess) is reacted with 1 ml
of hemoglobin solution for 30 min in an ice-water bath. Shake from time to
time. The reaction is terminated by passing the solution through a Sephadex
G-25 column equilibrated with 100 mM phosphate buffer pH 7.0. Fractions
of 0.5 ml are collected and used for the experiment.

2. Preparation of Spin-Labeled Cytochrome Oxidase

Three hundred μg of solid spin-label maleimide (tenfold excess) is reacted
with 1 ml of cytochrome oxidase solution for 15 min at room temperature.
The reaction is terminated by passing the solution through a Sephadex G-25
column equilibrated with 100 mM phosphate buffer pH 7.0, containing
0.5% Tween. The eluate is collected in 0.5 ml fractions. The fractions containing cytochrome oxidase are pooled and concentrated in a dialysis bag
covered with dry Sephadex.

3. Preparation of the Calibration Curve for Rotational Correlation Times Using Hemoglobin as a Standard

The 2nd harmonic 90° out-of-phase saturation transfer ESR spectrum of hemoglobin is measured at different temperatures and viscosities. The rotational correlation times are calculated from the Debye equation

$$\tau_c = \frac{4 \pi \eta r^3}{3 \, kT}$$

the radius of hemoglobin is 29 Å
the value of k is $1.38 \cdot 10^{-16}$ erg/K

Use the following glycerol-water mixtures and temperatures:

1. 40% glycerol   5°C   viscosity = $3.4 \cdot 10^{-2}$ Poise
2. 60% glycerol   5°C $3.7 \cdot 10^{-1}$ Poise
3. 80% glycerol   5°C   3.7   Poise
4. 90% glycerol $-12°C$ 81.1   Poise

The exact viscosities can be measured using a viscosimeter. The spectra obtained should look similar to the ones shown in Fig. 13.

τ_c (sec)

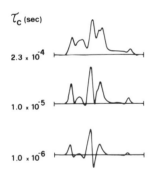

2.3 x 10^{-4}

1.0 x 10^{-5}

1.0 x 10^{-6}

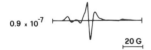

0.9 x 10^{-7}

20 G

Fig. 13. Second harmonic 90° out-of-phase saturation transfer ESR absorption spectra of spin-labeled hemoglobin. The conditions are as indicated in this section

From the spectra read the parameters shown in Fig. 14 and plot the ratios obtained against the calculated rotational correlation times.

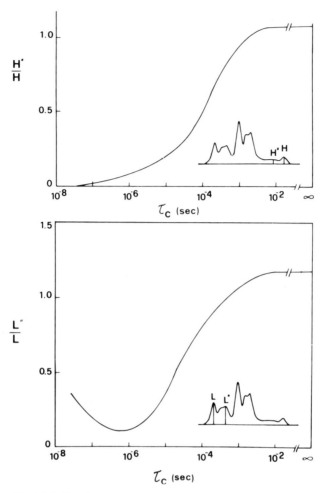

Fig. 14. Calibration curves prepared from data of saturation transfer ESR spectroscopy of spin-labeled hemoglobin

4. Measurement of the Rotational Correlation Time of Cytochrome Oxidase

After concentration measure the spectrum of the cytochrome oxidase sample in the ESR spectrometer at $4°C$. Read the same parameters from the spectrum as for hemoglobin and obtain the rotational correlation time from the curve.

Try to calculate the radius of the rotating cytochrome oxidase. Assume a viscosity of 1 to 10 Centipoise. The radius of the cytochrome oxidase monomer would be about 37 Å.

The whole experiment is done under the assumption that cytochrome c oxidase rotates isotropically like hemoglobin. Possibly this is not the case, but the results from these experiments are in fair agreement with the results obtained using other methods.

References

Berliner LJ (ed) (1976/79) Spin labeling: Theory and applications, vol I, 1976; vol II, 1979. Academic Press, London New York

Hubbell WL, McConnell HM (1971) Molecular motion in spin-labeled phospholipides in membranes. J Am Chem Soc 93:314–326

Hyde JS (1978) Saturation transfer spectroscopy. Methods Enzymol 49:480–511

Jost P, Griffith OH (1978) The spin-labeling technique. Methods Enzymol 49:369–418

Leute RK, Ullman EF, Goldstein A, Herzenberg LA (1972) Spin immunoassay technique for determination of morphine. Nature (London) New Biol 236:93–94

McCalley RC, Shimshick EJ, McConnell HM (1972) The effect of slow rotational motion on paramagnetic resonance spectra. Chem Phys Lett 13:115–119

Shimshick EJ, McConnell HM (1973) Lateral phase separation in phospholipid membranes. Biochemistry 12:2351–2360

Sigrist-Nelson K, Azzi A (1979) The proteolipid subunit of chloroplast adenosine triphosphatase complex. J Biol Chem 254:4470–4474

Stone TJ, Buckman T, Nordio PL, McConnell HM (1965) Spin-labeled biomolecules. Proc Natl Acad Sci USA 54:1010–1017

Thomas DD, Dalton LR, Hyde JS (1976) Rotational diffusion studied by passage saturation transfer electron paramagnetic resonance. J Chem Phys 65:3006–3024

Wertz JE, Bolton JR (1972) Electron spin resonance, elementary theory and practical applications. McGraw-Hill Book Company, New York

The Application of Fluorescence Spectroscopy to the Study of Biological Membranes

R.P. CASEY, M. THELEN, and A. AZZI

I. Introduction

Fluorescent techniques have had a dramatic impact in recent years on our knowledge of various complex biological systems and in particular of biological membranes. The measurement of fluorescence has the double advantage of being very sensitive and yet relatively easily carried out. In addition the technique is very versatile, insofar as it can provide information about molecular structure, mobility, orientation, and environment. The application of fluorescence to biochemistry has increased considerably since the introduction of extrinsic fluorescent probes, which can act as "reporters" of events and situations in otherwise nonfluorescent structures.

This introduction to the use of fluorescence will take the form of two practical exercises, one on "Fluorescence Techniques" in general and one on "Fluorescence Labeling". The protocols for these are described here, after the following short summary of relevant fluorescence theory. In these theoretical considerations, detailed derivation of formulae and lengthy discussions have been avoided. The student is advised to consult the cited works for more thorough presentations.

A. Basic Aspects of the Fluorescence Process

In most molecules at room temperature, the electrons are in the lowest vibrational energy level of the ground state (S_0 in Fig. 1). The absorption of

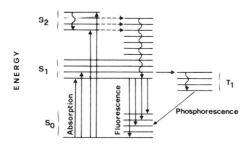

Fig. 1. The possible means by which the energy level of a molecule may alter on absorption of energy by electrons. *Upward directed lines* represents absorption of radiation and *downward directed lines* fluorescence. *Wavy lines* represent vibrational energy loss and *dotted lines* "intersystem crossover". For other details see text

light by such ground-state molecules provides the energy for the transition of some of the electrons to higher "excited singlet state" enery levels. Depending on the frequency of absorbed light, the electronic transition may be to any of the vibrational levels of the excited singlet states, as shown in Fig. 1. Owing to inter-molecular collisions, however, the molecules raised to upper vibrational levels of an excited state rapidly lose their excess energy and fall to the lowest vibrational level as indicated by the wavy lines of Fig. 1. In addition, through a process known as "internal conversion" molecules of a lower vibrational energy level of an upper excited state pass to a higher vibrational energy level of a lower excited state without loss of energy, as indicated by the dotted lines in Fig. 1. As the processes of internal conversion and collisional-relaxation between the excited states are more rapid than light emission, no fluorescence normally occurs from these upper excited-state situations. On the other hand, internal conversion from the first excited singlet state to the ground state is relatively slow, allowing the molecules to pass back from the lowest vibrational energy level of the first excited singlet state to any of the vibrational energy levels of the ground state, with the emission of light, i.e., *fluorescence*. The number of molecules which return to the ground state in this way instead of through other mechanisms determines the *Quantum Yield* of fluorescence. The transition between the lowest vibrational energy levels of S_0 and S_1 in both absorbance and emission is termed the "0-0 transition" and will be dealt with in more detail later.

A further mode of return from the first excited singlet state to ground-state is by "inter-system crossing" to the triplet (T_1) state. The subsequent transition to ground state with light emission (i.e., *phosphorescence*) is quantum mechanically "spin-forbidden" and does not normally occur at room temperature. Phosphorescence is dealt with in more detail in Parker (1968).

B. Fluorescence Excitation and Emission Spectra

Unlike the case for absorption spectroscopy, in fluorescence we deal with two kinds of spectrum; the excitation spectrum and the emission spectrum. In both, the intensity of emitted light is recorded as a function of wavelength. For the former, the wavelength at which emission is measured is kept constant and the wavelength of the excitation beam is varied; for the latter the converse applies. It is clear from Fig. 1 that, with the exception of the 0-0 transition, all peaks in the excitation spectrum will be of lower wavelength than those in the emission sepctrum (in fact the 0-0 bands are also separated in this way; see Section F for an explanation of this).

Figure 1 would indicate that the fluorescence spectra described above should be very uniform, consisting of regularly spaced, equally intense lines. Of course, Fig. 1 is a great over-simplification and this is not the case. The occurrence of a particular electronic transition depends on the "transition probability" which is dictated by the relative positions of the atoms of the molecule on light absorbance or emission. We can consider this most simply with the case for a diatomic molecule as shown in Fig. 2. There are drawn the energy profiles of the ground (ab) and excited (cd) state molecule and within each profile are shown the vibrational energy levels. Profile cd is shifted to slightly higher inter-atomic distances than ab, as the excited state electronic orbital is slightly larger than that in the ground state. Wave mechanics tells us that the most probable position for an atom is at the mid-point of the lowest ground-state energy level, i.e., the middle of line rs in Fig. 2. We are further told by the *Frank-Condon principle* that the most probable electronic transition is so rapid that it occurs in the absence of nuclear movement. This transition, which leads to the main band in the excitation spectrum, can thus be described by the vertical line tv in Fig. 2. The molecule then rapidly relaxes to the lowest vibrational energy level (xy) and again the Franck-Condon principle dictates that the most likely transition back to the ground state (leading to the most intense emission band) is given by the line zq.

This simple example shows how physically preferred electronic transitions will dictate the shapes of the fluorescence spectra. When one considers that in the system discussed other transitions, while less likely, are possible, and that all the molecules which we will deal with are much more complicated than that considered above, it becomes clear how more complex fluorescence spectra arise.

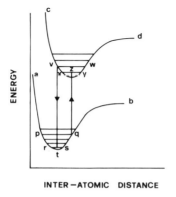

INTER–ATOMIC DISTANCE

Fig. 2. Energetic description of the absorption and fluorescence processes. The lines *ab* and *cd* represent the energy profiles of the ground and first excited state molecules respectively. The *vertical lines* represent vibrational energy levels. Other details are described in text

C. Fluorescence Life-Time

One of the most informative parameters in fluorescence spectroscopy is the *life-time* of the excited state. This is defined as follows: If a number of molecules, n_0, are excited to a higher level and can only return to ground state through fluorescence, then the number emitting light in unit time is given by the equation:

$$\frac{dn}{dt} = -k_F n \tag{1}$$

where n is the number of excited molecules present at time t and k_F is the first order rate constant for fluorescence emission.

Integration of (1) gives:

$$n = n_0 e^{-k_F t} \tag{2}$$

The rate of light emission Q is then given by:

$$Q = \frac{-dn}{dt} = k n_0 e^{-k_F t} = Q_0 e^{-k_F t} \tag{3}$$

The "natural" fluorescence life-time, τ_0, is then defined as

$$\tau_0 = \frac{1}{k_F} \tag{4}$$

i.e., it is the time taken for the intensity (and thus the population of excited molecules) to fall to 1/e of its initial value. As the molecule can normally return to ground state through other means than only fluorescence, the real life-time, τ, is nearly always less than τ_0 and is defined as the inverse of the sum of the rates of *all* the processes, whereby the excited molecule can return from the lowest excited singlet state to the ground state.

The natural and real fluorescence life-times are related by the expression:

$$\phi = \frac{\tau}{\tau_0} \tag{5}$$

where ϕ is the quantum yield.

For a more detailed treatment of the theory and measurement of life-times the student should consult Parker (1968) and Azzi (1975).

D. Quenching of Fluorescence and Energy Transfer

The quantum yield of fluorescence can be defined as:

$$\phi = \frac{k_F}{k_F + k_T + k_I + k_Q [Q]} \tag{6}$$

where the k's are the first order rate constants for fluorescent emission, (k_F); crossover to the triplet state, (k_T); internal conversion to ground state (k_I) and relaxation through interaction with "quenching" molecules (k_Q), having concentration $[Q]$. Whilst *static quenching* may occur through formation of a nonfluorescent complex of a quencher with the fluorochrome, *dynamic quenching* is of more interest to us. Here, energy is nonradiatively transferred, from a donor molecule in the first excited singlet state to a neighboring ground-state acceptor molecule, causing quenching of the fluorescence of the former and excitation of the latter. This leads to the particular useful technique of "distance measurement through energy transfer", which can be used in biological systems at both the inter-molecular and intra-molecular levels. The transfer efficiency, T, (i.e., the ratio of the quanta emitted by the donor in the presence to that in the absence of the acceptor) and the distance between the two chromophores, R are related by:

$$\frac{T}{1 - T} = \left[\frac{R_0}{R} \right]^6 \tag{7}$$

where R_0 is the value of R when the rate of energy transfer is equal to the sum of the rates of all competing processes. R_0 is calculated using the Förster (1951) equation (3)in its corrected form (Latt et al. 1965).

$$R_0 = \left[\frac{9000 \, l_n \, 10 \, k^2 \phi}{128 \, \pi^5 \, n^4 \, N} J \right]^{1/6} \tag{8}$$

The experimentally determined values here are ϕ, the quantum yield of donor fluorescence in the absence of energy transfer, and J, the integral of overlap of the donor emission and acceptor excitation spectra. N is Avogadro's Number, n is the refractive index of the surrounding medium and k^2 is a factor (normally assumed) which represents the relative orientations of the chromophores under consideration. A more extensive discussion of energy transfer is given in Stryer (1978).

E. Fluorescence Polarization

By irradiating the fluorescent species with plane-polarized light and measuring changes in the polarization of light on emission, considerable information can be obtained concerning the orientation and mobility of biological molecules and the viscosity of their surroundings. When discussing fluorescence polarization it facilitates matters to consider the fluorophore as an electronic oscillator having a transition moment in the direction of the electronic transition. The degree of polarization, p, of the emitted beam is measured at right angles to the plane-polarized excitation beam and is defined as follows (Weber 1953).

$$p = \frac{I_{\parallel} - I_{\perp}}{I_{\parallel} + I_{\perp}} \tag{9}$$

where $I_{\parallel}$ and $I_{\perp}$ are the intensities of the emitted beam when viewed through a polarized filter oriented parallel and perpendicular to the plane of polarization of the excitation beam. If the fluorescent molecules are randomly oriented but rigidly held (i.e., they do not change their orientation between absorption and emission), then p takes the value p_0, which is given by

$$p_0 = \frac{3 \cos^2 \theta - 1}{\cos^2 \theta + 3} \tag{10}$$

where θ is the angle between the transition moments of the absorption and emission processes (in this case $\theta = 0$, thus $p_0 = 0.5$).

In the case of our experiments the molecules are not rigidly held and molecular motion occurring between absorption and emission will alter the degree of polarization such that:

$$\frac{1}{p_0} - \frac{1^{\cdot}}{3} = \left[\frac{1}{p} - \frac{1}{3} \right] \left[1 + \frac{3\tau}{\rho} \right] \tag{11}$$

this is the Perrin equation (1926). Here, ρ is the rotational relaxation time and τ is the fluorescence life-time. For a spherical molecule, ρ and the viscosity of the medium, η, are related by

$$\rho = \frac{3 V \eta}{RT} \tag{12}$$

where V is the molar volume.

Another informative application of fluorescence polarization is the measurement of the polarization spectrum. There, the degree of polarization of the emitted light is plotted as a function of the wavelength of the excitation beam whilst measuring at a constant wavelength. The polarization of emitted light at a particular wavelength depends on the relative orientations of the transition moments of the absorption and emission process. The polarization spectrum thus allows detection of electronic transitions giving overlapping spectra but having differently oriented transition vectors.

It also allows the determination of the relative orientations of the transition moments associated with the various absorption bands and those with the emission band from the lowest excited state.

A more comprehensive account of fluorescence polarization, including events occurring at the electronic and molecular levels, is given in Parker (1968), Förster (1951) and Shinitzky and Barenholz (1978).

F. Environmental Effects on Fluorescence Properties

The possible effects of molecular environment on fluorescence are many and varied. This has the mixed blessing that fluorescence measurements can tell us much about the surroundings of the fluorescent species and that such measurements can be easily misinterpreted. For a full coverage of these effects the reader should consult Azzi (1975) and Radda (1971). Here, however, we shall only discuss briefly one of these: the effect of medium polarity.

The degree of separation of the absorbance and fluorescent emission bands which should otherwise overlap at the 0-0 transition depends on the polarity of the medium. This is because dipolar interaction of the fluorescent molecules with dipoles of surrounding solvent molecules determines their extent of solvation. The change in the dipole moment of the fluorochrome on excitation means that a change in its now unstable solvated state occurring before emission will cause a decrease in its energy and therefore emission will occur at a longer wavelength than excitation. Lippert (1957) has shown that this separation of excitation and emission maxima is greater the more polar the solvent. This has the consequence of the well-known "blue-shift" in the emission maxima of many fluorescent molecules on changing from a polar to a nonpolar environment.

Another interesting effect is the increased quantum yield of some substances, e.g., anilinonaphthalene sulfonate (ANS), on passing from a polar to a nonpolar environment. It has been suggested (Brand and Gohlke 1972) that the energy difference between the first-excited singlet (S_1) and triplet (T_1) states is greater in nonpolar environments as there is less stabilization of S_1 through solvent relaxation than in polar environments. Thus, the rate of intersystem crossing is reduced in nonpolar media and the quantum yield of fluorescence increased.

G. Derivation of the Lateral Diffusion Coefficient from Measurements of Excimer Formation

While this is a specialized area of fluorescence it is one particularly useful in biomembrane research. If molecular diffusion is considered as a "random-jump" process, the diffusion coefficient (D) is related to the number of molecular collisions per second (ν_{col}), the length of one diffusional jump (λ), the van der Waals diameter (d_c) and the concentration of diffusing molecule in units of mol/Å^2 (c) by:

$$D = \frac{\nu_{col}\lambda}{4\,d_c c} \tag{13}$$

If the molecules undergo a reversible physical reaction on collision, ν_{col} is then proportional to the second-order rate constant of the reaction and D can be derived, provided the reaction can be monitored. A reaction which lends itself to this type of study particularly well is the interaction between a fluorescent molecule in the ground state (A) with an excited molecule (A*) to form an excited dimer or excimer (AA*). The following formula has been derived (12) relating the diffusion coefficient to measurable fluorescence parameters of A* and AA*.

$$D = \frac{I'}{\kappa I}\,\frac{1}{40\,\tau c} \tag{14}$$

where I and I' are the intensity of emission of A* and AA* respectively, κ is a proportionality coefficient expressing the relative spectral distribution of the monomer and excimer emissions and τ is the fluorescence life-time of the excimer.

II. Apparatus and Materials

A. Apparatus

Perkin-Elmer MPF-2A Fluorescence Spectrophotometer having excitation and emission beams with independently and continuously variable wavelengths and slit-widths. The cuvette holder is thermostated with a water bath
polarizing filters for the fluorometer
MSE sonication apparatus with nitrogen input above the sample holder

quartz fluorescence cuvettes, 3 ml capacity
1.5 and 10 ml glass pipettes
25 and 50 μl microsyringes
vessels with ground glass stoppers for the volatile solvents
glass test tubes
plastic vessel for the sonication
plastic cuvette stirrer

B. Materials

recrystallized soybean phospholipids (asolectin)
porphyrin cytochrome c (prepared according to Vanderkooi and Erecinska
    1975)
beef heart mitochondria
1 mM-8-anilino-1-naphthalenesulfonate (ANS) in ethanol
10 mM-pyrene in ethanol, acetone and cyclohexane
10 mM 9-aminoacridine in ethanol
0.2 mM-valinomycin in ethanol
1 mM-carbonyl cyanine m-chlorophenylhydrazone (CCCP) in ethanol
ethanol and octanol (analytical grade)
298 mM-sucrose, 1 mM-KCl, 10 mM Hepes, pH 7.4
150 mM-KCl, 10 mM-Hepes, pH 7.4
15 mM-KCl
150 mM-KCl
225 mM-mannitol, 75 mM-sucrose, 1 mM-EDTA
10 mM-sodium phosphate buffers at pH 1.3 and 7.0
200 mM-sucrose, 20 mM-Hepes, pH 7.0
6 M-guanidinium chloride
90% glycerol
1% Tween 80, 20 mM potassium phosphate, pH 7.0
doubly distilled water

III. Experimental Procedures

A. Fluorescence Techniques

To prepare asolectin monolayer vesicles, sonicate a mixture of 10 mg of
asolectin in 3 ml of 150 mM-KCl, 10 mM-Hepes, pH 7.4 for 2 min on ice at
4 μ power. Add to this 7 ml of 150 mM-KCl, 10 mM-Hepes, pH 7.4. To pre-

pare asolectin multilayer vesicles having low and high internal K^+ concentrations mix 20 mg of asolectin in 1 ml of 298 mM-sucrose, 1 mM-KCl, 10 mM-Hepes, pH 7.4 or 1 ml of 150 mM-KCl, 10 mM-Hepes, pH 7.4 at full speed for 1 min on a vortex mixer. Then centrifuge the suspension for 5 min at full speed on a bench centrifuge to remove lipid aggregates.

1. Experiments Using ANS

These experiments illustrate the environmental effects on fluorescence described in Section F of the Introduction. The effects of environmental polarity on fluorescence can be observed by measuring the fluorescence of ANS in various solvents, in the presence of membrane vesicles, and with vesicles which possess a transmembrane electrical potential difference. The last of these observations results from additional partitioning of ANS into the membrane lipid in response to the membrane potential because of the negative charge on the probe.

a) Effects of Solvent Polarity on Fluorescence Properties of ANS. Prepare the following ethanol/water mixtures

	Ethanol	Water
A	2.5 ml	—
B	1.5 ml	1 ml
C	1 ml	1.5 ml
D	—	2.5 ml

Add 20 μl of 1 mM-ANS to each of these and to 2.5 ml of octanol. Run emission spectra from 390 nm to 540 nm and excitation spectra from 260 nm to 420 nm on A to D (initially using 380 nm as the excitation wavelength). A clear red shift in the maximal emission wavelength and decrease in the quantum yield (i.e., intensity of fluorescence) should be observed as the medium polarity increases. Repeat these measurements using the octanol sample. As octanol is less polar than ethanol the maximal emission wavelength should be more blue-shifted and the quantum yield higher than that with ethanol.

b) Fluorescence Properties of ANS in the Presence of Membrane Vesicles. 50 μl of asolectin multilayer vesicles with a low internal K^+ concentration (1 mM) are added to 2.5 ml of 150 mM-KCl, 10 mM-Hepes, pH 7.4. Measure the emission spectrum of the sample from 400 nm to 520 nm with excitation at 372 nm in the presence and absence of 40 μl of 1 mM-ANS. Comparison of the difference between these spectra (i.e., the fluorescence of ANS in the presence of membranes) with the emission spectrum of ANS in medium only should show a blue shift in fluorescence emission maximum and increase of the quantum yield, indicating that the polarity of the ANS environment in the membrane is about the same as that of octanol.

Add a further 50 μl of the above vesicles to 2.5 ml of 150 mM-KCl, 10 mM-Hepes, pH 7.4 followed by 40 μl of 1 mM-ANS. Follow the fluorescence emission at 468 nm with excitation at 380 nm. Add 5 μl of 0.2 mM-valinomycin. The fluorescence should rise immediately and then slowly decay, in response to the transmembrane electrical potential triggered by valinomycin addition. When the signal has decayed to about 75% of its initial level add 25 μl of 1 mM-CCCP. This substance is a protonophore which collapses the membrane potential and the ANS fluorescence should now decay much more rapidly.

2. Detection of a Transmembrane pH Difference Using 9-Amino Acridine

9-Amino acridine is a weak base. As such it distributes across a transmembrane pH difference (inside acidic) causing it to concentrate inside the vesicles. It is also fluorescent and when concentrated its fluorescence emission is quenched by static quenching as described in Section D above. Here we use an artificially imposed pH difference in multilayer vesicles to demonstrate this.

50 μl of multilayer vesicles with a high internal K^+ concentration (150 mM) are added to 2.5 ml of 298 mM-sucrose, 1 mM-KCl, 10 mM-Hepes, pH 7.4. 5 μl of 10 mM-9-amino acridine is added to this and the fluorescence at 430 nm observed with excitation at 400 nm. 5 μl of 0.2 mM-valinomycin is added and the fluorescence should drop slowly to a final steady level, according to the above rationale. Repeat this experiment with 10 μl of 1 mM-CCCP present from the beginning. The full valinomycin-induced drop in fluorescence should now occur immediately. This experiment also gives a qualitative indication of the resistance of the membrane to proton movement.

3. Measurements of Pyrene Excimer Formation

In these experiments the proportion of pyrene molecules present as fluorescent dimers (excimers) under a number of conditions is measured, providing quantitative information about the mobility of the probe.

a) **Determination of the Viscosity Coefficient (η) of the Medium.** Here η is calculated from the ratio of pyrene monomer and excimer fluorescence using the following formula:

$$\eta = \frac{Q}{Y}\left(\frac{I_M}{I_D} \cdot c \cdot T\right)$$

Where I_M and I_D are the fluorescence emission of the monomer and dimer forms of pyrene respectively, c is the molar concentration and T is the absolute temperature. Y is the sum of a number of rate constants in the fluorescence process, and has the following values:

    for acetone, 35×10^6
    for ethanol, 27×10^6
    for cyclohexane, 22×10^6

Q is a constant of proportionality determined from τ measurements and has the following values:

    for acetone, 4.0×10^6
    for ethanol, 13.5×10^6
    for cyclohexane, 15.7×10^6

Measure I_M and I_D for 10 mM solutions of pyrene in ethanol, acetone, and cyclohexane at 480 nm using excitation at 360 nm. From these calculate η for these solvents.

b) **Measurement of the Diffusion Coefficient of Pyrene in Phospholipid Membranes at Various Temperatures.** Here we make use of the theory discussed in Section G. We shall assume the following values for the parameters of Eq. (14): $\kappa = 0.8$ and $\tau_0 =$

    100 ns at 45°C
    115 ns at 35°C
    130 ns at 25°C

The concentration of pyrene molecules in molecules. $Å^{-2}$ (c) we calculate as

$$c = \frac{R}{F}$$

where R is the molar ratio of pyrene:lipid and F is the surface area of a pyrene molecule (assumed to be 58 $Å^2$).

Add 34 μl of 10 mM-pyrene in ethanol to 10 ml of asolectin monolayer vesicles prepared as described above. To decrease the concentration of O_2 which quenches pyrene fluorescence bubble the solution with nitrogen for

10 min. Transfer 2.5 ml of this mixture as rapidly as possible to a cuvette and then bubble again with nitrogen for a few seconds and then seal the cuvette (e.g., with parafilm). Run an emission spectrum of the sample at 25°C from 360 nm to 520 nm with excitation at 320 nm. Measure the fluorescence intensity from the pyrene monomer (at 385 nm) and the dimer (at 465 nm). Repeat these measurements after adjusting the temperature to 35°C and 45°C. Calculate the diffusion coefficients of pyrene in the lipid bilayer at these temperatures using the theory outlined above.

N.B. The values obtained for D will be in the units of a second order rate constant, i.e., 1 mol^{-1} s^{-1}. To compare with the values reported by Galla and Sackmann (1974) these must be converted to the units normally used for a diffusion coefficient (molecule $^{-1}$ cm^2 s^{-1}) by dividing by 1.34×10^{14}.

B. Fluorescence Labeling

Beef heart mitochondria are depleted of cytochrome c as follows (Boveris et al. 1972). The mitochondria (suspended in 0.015 M-KCl) are incubated at 0°C for 30 min with occasional stirring. They are then centrifuged at 8000 g for 10 min at 4°C and the sediment is washed twice in 0.15 M-KCl and once in 0.225 M-mannitol, 0.075 M-sucrose, 1 mM-EDTA and finally suspended in this medium.

1. Add 30µl of 0.3 mM porphyrin cytochrome c to 3 ml aliquots of 10 mM-sodium phsopahte at pH 7.0 and 1.3. Measure the excitation (from 320 to 620 nm) and emission (from 560 to 720 nm) spectra of this substance at these pH values. Consider possible reasons for the differences in the spectra.

2. Add 10 µl of 0.3 mM porphyrin cytochrome c to 3 ml of either 10 mM-sodium phosphate, pH 7.0 or 6 M-guanidinium chloride. Measure the emission spectra of the samples from 290 to 670 nm with excitation at 285 nm. Energy transfer from tryptophan to the porphyrin system (see Sect. D) should be seen in the absence, but not in the presence, of guanidinium chloride, owing to the denaturing effects of this substance.

3. Add 30 µl of 30 µM porphyrin cytochrome c to 3 ml aliquots of 10 mM-sodium phosphate pH 7.0 or 90% glycerol. Measure the polarized excitation spectrum of the probe in these solvents from 460 to 660 nm measuring emission at 620 nm. This is done by measuring fluorescence at each wavelength setting using the polarizing filters under the following conditions (where the filter of the excitation beam is the polarizer and that of the emission beam is the analyzer):

analyzer vertical, polarizer vertical $(I_\parallel)$

analyzer horizontal, polarizer vertical $(I_\perp)$

analyzer vertical, polarizer horizontal $(I_\perp{}^0)$

analyzer horizontal, polarizer horizontal $(I_\parallel{}^0)$

$I_\parallel$ and $I_\perp$ are defined for Eq. (9). $I_\parallel{}^0$ and $I_\perp{}^0$ should be equal and if they are not then they represent an intrinsic polarization of the system and thus the quantity $I_\perp$ has to be corrected by multiplying by $\left(\dfrac{I_\perp{}^0}{I_\parallel{}^0}\right)$. Thus,

$$p = \frac{I_\parallel - \left(I_\perp \times \dfrac{I_\perp{}^0}{I_\parallel{}^0}\right)}{I_\parallel + \left(I_\perp \times \dfrac{I_\perp{}^0}{I_\parallel{}^0}\right)}$$

The polarization should be much more pronounced in the glycerol medium.

4. The fluorescence emission spectrum of porphyrin cytochrome c overlaps in part with the absorbance spectrum of cytochrome c oxidase. Consequently, cytochrome c oxidase quenches the fluorescence of porphyrin cytochrome c and the distance between the absorptive groups of the oxidase (i.e., the hemes) and the fluorescent group of the porphyrin cytochrome c (i.e., the porphyrin ring) can be calculated from the maximal quenching of fluorescence by the rationale outlined in Section D (Dockter et al. 1978).

Add 11 μl of 30 μM porphyrin cytochrome c to 3 ml of 20 mM-potassium phosphate, 1% Tween 80, pH 7.0. Measure the fluorescence emission at 620 nm, with excitation at 350 nm. Add cytochrome oxidase from a 50 μM solution to give final concentrations of 0.1, 0.2, 0.6, and 0.8 μM. Measure the change in fluorescence on cytochrome oxidase addition and draw a graph of the reciprocal of the fluorescence quenching vs the reciprocal of the cytochrome oxidase concentration. By extrapolating to zero on the x-axis (i.e., infinite cytochrome oxidase concentration) determine the maximal quenching of fluorescence.

Calculate R of Eq. (7) for this system using the value of T determined as above and the following values for the other parameters (Dockter et al. 1978): $\phi = 0.014$, $k^2 = 1.5$, $J = 0.235 \times 10^{-13}$ cm^3 M^{-1}, n = 1.4.

5. Add 10 μl of 0.3 mM porphyrin cytochrome c to 3 ml of 10 mM-sodium phosphate pH 7.0. Measure the plarization of fluorescence of the probe at 620 nm with excitation at 500 nm as described above. Repeat these measurements with cytochrome c-depleted mitochondria added to give a final concentration of 3 mg/ml. The fluorescence polarization should be increased

when porphyrin cytochrome c binds to the mitochondria, indicating increased immobilization of this substance when bound.

References

Azzi A (1975) Q Rev Biophys 8:237–316
Boveris A, Erecinska M, Wagner M (1972) Biochim Biophys Acta 256:223–242
Brand L, Gohlke JR (9172) Annu Rev Biochem 41:843–868
Dockter ME, Steinemann A, Schatz G (1978) J Biol Chem 253:311–317
Förster T (1951) Fluoreszenz organischer Verbindung. Vandenhoek & Ruprecht, Göttingen
Galla H-J, Sackmann E (1974) Biochim Biophys Acta 339:103–115
Latt SA, Cheung HT, Blout ER (1965) J Am Chem Soc 87:995–1003
Lippert E (1957) Z Elektrochem 61:962–975
Parker CA (1968) Photoluminescence of solutions. Elsevier, Amsterdam
Perrin F (1926) J Phys Radium 7:390–401
Radda GK (1971) Curr Top Bioenerg 4:81–126
Shinitzky M, Barenholz Y (1978) Biochim Biophys Acta 515:367–394
Stryer L (1978) Annu Rev Biochem 47:819–846
Vanderkooi JM, Erecinska M (1975) Eur J Biochem 60:199–207
Weber G (1953) Adv Protein Chem 8:415–459
Yu C, Yu L, King TE (1975) J Biol Chem 250:1383–1392

Use of a Potentiometric Cyanine Dye in the Study of Reconstituted Membrane Proteins

H. LÜDI, H. OETLIKER, and U. BRODBECK

I. Introduction and Aims

The reconstitution procedure described on p. 103 of this book yields phospholipid vesicles with an outside diameter of 25–30 nm. The membrane protein of interest is contained in a single bilayer structure, which separates a defined inside volume from the suspension medium. An advantage of small vesicles is the fact that the membrane protein is inserted in a non-random fashion, e.g., more than 85% of the α-bungarotoxin-binding sites are oriented to the outside of the vesicles (see chapter by Lüdi and Brodbeck, this vol.). On the other hand, the small vesicular volume and the very small ratio of intravesicular space to extravesicular space makes detection of rapid ion movements rather difficult with conventional techniques (tracer fluxes, detection of concentration changes by means of ion-sensitive electrodes etc.).

Under certain conditions these difficulties can be overcome by the use of a different detection method.

If the ion fluxes occur in an electrogenic manner they can be detected as changes in membrane potential and one does not need to measure changes in ion-concentration.

Since the membrane capacitance of such vesicles is small, due to their small surface area, very little charge transfer is sufficient to produce substantial changes in membrane potential of these vesicles.

Changes in membrane potential of structures too small to be impaled by a microelectrode can be measured by the use of "potentiometric dyes" (see Waggoner 1979 for review).

In this experiment changes in membrane potential of phospholipid vesicles are followed by monitoring the fluorescence of indodicarbocyanine which was added to the suspension medium.

Phospholipid vesicles formed in 200 mM KCl, diluted into 200 mM NaCl solution and made permeable for K^+ by valinomycin are used as control vesicles. Their change in fluorescence upon a standard increase in the outside potassium concentration is compared with the fluorescence response of vesicles containing the purified acetylcholine receptor protein beside valinomycin.

Such a method combines the advantage of a nonrandom insertion of the membrane protein with the detection of protein-mediated changes in membrane permeability. The time resolution of the experimental layout, described here, is in the order of 1–3 s, it could be increased to the ms-range by the use of fast mixing and recording techniques.

A wide variety of membrane proteins can be studied by this method, provided that changes in permeability or pumping activity result in a change in transmembrane potential. Electrogenicity can sometimes be reached by appropriate selection of permeable or impermeable ions in the vesicles forming solution or the suspension medium.

II. Theory

A. Detection Limit

The difference in detection limit achievable with the "potentiometric dye method" versus the resolution reached by "conventional" methods is illustrated by the following rough calculation. It is based on data given by Huang and Mason (1978) and the following assumptions.

Formation of vesicles: 1 mg phospholipid/ml buffer, containing 200 mM KCl and 10 mM Tris-HCl, pH 7.4. Density of lipid membrane, including structural water: $\sim$ 1.0. Thickness of lipid bilayer including the inner and outer hydration shell: 5 nm (Huang and Mason 1978). Outside diameter of vesicles: 30 nm. As generally assumed for biological membranes, the specific membrane capacitance is taken as 1 $\mu F/cm^2$.

For simplicity the calculation is based on "control vesicles" made specifically permeable for K^+ by valinomycin [0.03 $\mu g/ml$ (Mueller and Rudin 1967)] which do not contain membrane proteins.

Based on the above assumptions an average vesicle has a surface area of 2827 nm^2 and encloses a volume of 4189 nm^3 which contains $\sim$ 500 K^+ ions and $\sim$ 500 Cl^- ions. The membrane volume amounts to 9948 nm^3. 1 mg of phospholipids therefore forms $\sim 10^{14}$ vesicles if an efficiency of 100% is asssumed; the total volume enclosed by 1 mg of phospholipid is 4.21×10^{17} nm^3.

The ratio of inside to outside volume in the solution where vesicles are formed is 1:2400.

An ion distribution similar to the situation in a living cell, as far as potassium is concerned, can be achieved by diluting 100 μl of vesicles into 3 ml of (K^+ free) 200 mM NaCl, 10 mM Tris-HCl buffer solution. This results in a composition of the suspension medium of 193.5 mM NaCl and 6.5 mM KCl. The ratio inside volume/outside volume is reduced to $\sim$ 1:74,000.

Vesicles made specifically permeable for potassium ions by valinomycin will ideally lose potassium until their membrane potential reaches the value predicted by the Nernst equation, upon dilution into a NaCl medium.

$$E_M = -59 \log \frac{[K]_i}{[K]_o} \tag{1}$$

Where E_M is the membrane potential and $[K^+]_i$ and $[K^+]_o$ are the potassium concentrations (more accurate activities) inside and outside of the vesicles. The numerical value of this equation under the conditions described above is -87 mV (inside negative).

To charge the surface membrane to that potential an electrogenic flow of only 15 K^+ ions per vesicle is needed. This reduces the number of potassium ions on the inside by 3% and increases the outside potassium concentration by 17 nM or from 6.500'000 to 6.500'017 mM or by a factor of $(1 + 3 \times 10^{-6})$. Such small concentration changes are clearly below the detection limit of "conventional methods".

A change in membrane potential, on the other hand, can easily be detected down to less than 10 mV (this corresponds to an average electrogenic loss of $1-2$ K^+ ions per vesicle) by optical methods, depending on the dye and the fluorimeter used (see Fig. 2).

B. Mechanism for Potential Dependent Optical Changes

It has been suggested that the potential dependent fluorescence of cyanine dyes is mainly due to a potential dependent partition of the dye between the aqueous phase of the suspension medium and the hydrophobic environment of the lipid membrane. The fluorescence characteristics of these dyes are highly sensitive to the degree of polarity of their microenvironment. See Sims et al. (1974) and Waggoner (1979).

C. Calibration of Optical Response

A calibration curve (calculated membrane potential versus fluorescence intensity) can be obtained by measuring the fluorescence at different outside potassium concentrations. See Waggoner (1979) and Hoffmann and Laris (1974).

Experiments performed on neurons, where the membrane potential was measured simultaneously by an intracellular electrode, showed good agreement between the electrical and the optical determinations. See Cohen and Salzberg (1978) for review.

D. Permeabilities Induced by Membrane Proteins

Information about additional permeabilities induced by membrane proteins can be obtained by comparing the fluorescence responses of vesicles containing only valinomycin with the fluorescence responses from vesicles containing the membrane protein in addition to valinomycin upon a series of changes in the outside potassium concentration.

The membrane potential of vesicles being permeable for different monovalent ions such as K^+, Na^+, Cl^- can be expected to follow the constant field equation as given by Hodgkin and Katz (1949).

$$E_M = \frac{RT}{F} \ln \frac{P_{Na}[Na]_o + P_K[K]_o + P_{Cl}[Cl]_i}{P_{Na}[Na]_i + P_K[K]_i + P_{Cl}[Cl]_o} \qquad (2)$$

Where R = gas constant; T = absolute temperature; F = Faraday's constant; P_{Na}, P_K, P_{Cl} = the permeability constants for Na, K and Cl; $[Na]_o$. $[K]_o$, $[Cl]_o$ = Na, K, Cl concentrations (more strictly activities) of the medium or of the volume inside the vesicles respectively (i). The components of the buffers are not listed in the formula because their permeability and therefore their contribution to the membrane potential in general is small compared to the other ions.

E. Potential Independent Changes of Fluorescence

Under appropriate conditions "potentiometric dyes" are reliable indicators of changes in membrane potential. One has, however, to be aware of the possibility that fluorescence responses might be involved which are not due to changes in membrane potential.

Such membrane potential-independent changes in dye fluorescence have been reported from work done with the purified sarcoplasmic reticulum ATPase protein, which exhibits a calcium concentration-dependent fluorescence when it is stained with cyanine dyes. See Russel et al. (1979) and Oetliker (1980).

In order to know the extent of such potential-independent fluorescence contributions it is advisable to make appropriate control measurements such as check if the protein of interest alone produces changes in fluorescence upon changing the ionic composition of the suspension medium. Do the fluorescence responses persist after the vesicles have been made unspecifically permeable by other ionophores or destroyed with detergents?

III. Equipment and Solutions

egg phosphatidyl choline vesicles in 0.2 M KCl, 10 mM Tris-HCl, pH 7.4

1 mg phospholipid/ml, containing the purified acetylcholine receptor (prepared as described on p. 70) and same vesicles but prepared without protein, as control vesicles

Valinomycin (Sigma & Co.), 11 μg/ml in ethanol

dye (1,3,3,1',3',3',-hexamethylindodicarbocyanine, NK 529, from Nippon Kankoh Shikiso Kenkyusho, Okayama, Japan, 50 μg/ml in ethanol:water (1:10)

[125I]-α-bungarotoxin (acc. to Fulpius et al., see Chap. 2.3)

0.2 M KCl, 10 mM Tris-HCl, pH 7.4

0.2 M NaCl, 10 mM Tris-HCl, pH 7.4

3 M KCl

3 M NaCl

Perkin Elmer MPF-3L fluorescence photospectrometer: excitation = 620 ± 20 nm, emission = 680 ± 20 nm

3 ml fluorescence cuvettes (glass or plastic)

facilities for permanent and regular stirring

IV. Experimental

A. Determination of Excitation and Emission Parameters

To be able to optimize the time-resolved optical responses produced by changes in membrane potential, it is of interest to know the absorption spectrum and the fluorescence emission spectrum for the particular dye in the respective solution.

Generally it is advisable to use rather low dye concentrations to avoid saturation of presumptive dye binding sites in the vesicular membranes and to avoid absorption of the emitted fluorescence by the dye itself.

A good first choice for the excitation wavelength is a rather broad band near the absorption peak. It is worth while to open the excitation slit as wide as possible, this leads to a drastic increase in light intensity and a proportional increase in signal size; the noise is expected to rise only with the square root of light intensity; this leads therefore to an improved signal-to-noise ratio.

Comparison of the emission spectra under "resting" conditions and conditions where the membrane potential is expected to be deflected (increased

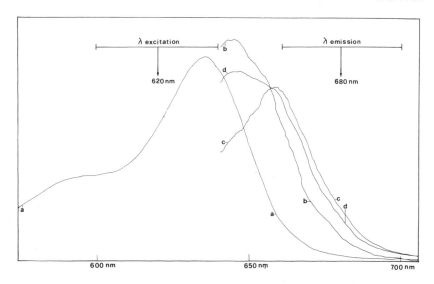

Fig. 1. Relative spectra of indodicarbocyanine. 0.5 μg indodicarbocyanine have been diluted in 3 ml 200 mM NaCl 10 mM Tris-HCl buffer solution, pH 7.4. *Trace a* absorption spectrum. Emission spectra obtained with excitation wavelength 620 ± 20 nm and slit width of 4 nm for scan from 640 to 710 nm. *Trace b* same solution as in *a; trace c* after additon of 100 μl protein-free phospholipid vesicles and 10 μl valinomycin solution; *trace d* after addition of 5 μl 3 M KCl solution

$[K^+]_o$) will help to select a band in the emission spectrum where the relative change in light intensity is maximal for a given change in membrane potential.

Figure 1 shows an example of such spectral determinations for an indodicarbocyanine dye. Trace a represents an absorption spectrum of a solution containing 0.5 μg dye in 3 ml 200 mM NaCl, 10 mM Tris-HCl buffer without vesicles.

The emission spectra b–d were measured while the dye was excited in a wavelength range from 600 to 640 nm, by scanning the wavelength range from 630 to 710 nm with an emission slit width of 4 nm. Spectrum b is obtained from dye alone in buffer solution, trace c is recorded after 100 μl of protein-free phospholipid vesicles containing valinomycin are added to the cuvette. According to Eq. (1) the vesicular membrane potential is expected to be − 87 mV. Trace d is recorded after 5 μl of 3 M KCl have been added, this is expected to shift the membrane potential to − 72 mV. It is obvious that the blue shift of the emission spectrum results in a decrease in light intensity when the emitted light in the range from 660 to 700 nm is used to monitor time-resolved changes in fluorescence. It is advisable to perform all measurements under constant and regular stirring of the cuvette content to avoid drifts in fluorescence.

B. Measurements of Changes in Membrane Potential in Phospholipid Vesicles

Figure 2 trace b shows an example of calibration measurements for estimating changes in membrane potential of phospholipid vesicles by monitoring indodicarbocyanine fluorescence. Since the K^+-permeability, induced by valinomycin, is the dominent permeability in those vesicles, see Mueller and Rudin (1967), the membrane potential is expected to vary linearly with $-\log [K]_o$. $[K]_i$ can be assumed to remain constant for practical purposes).

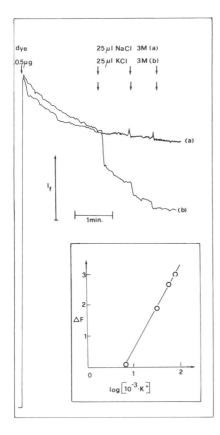

Fig. 2. Fluorescence responses of indodicarbocyanine upon changes in the ionic composition of the suspension medium. 100 μl phospholipid vesicles, containing 200 mM KCl 10 mM Tris HCl buffer are diluted into 3 ml 200 mM NaCl, 10 mM Tris-HCl buffer solution. *Trace a* shows relative fluorescence upon adding 25 μl 3 M NaCl at *arrows*. The NaCl concentration is increased from 200 to 275 mM in three steps by this. *Trace b* shows the decrease in fluorescence caused by the addition of 25 μl of 3 M KCl at *arrows* in a different sample. The outside KCl concentration has been raised from 6.5 to ~82 mM. *Inset* change of fluorescence (ΔF) upon 25 μl 3 M KCl plotted as function of the log of the K-concentration in the suspension medium. cf. Eq. (1). I_f Fluorescence intensity. *Bar* represents 20%

The inset in Fig. 2 shows that the fluorescence varies linearly with $\log [K]_o$. It can therefore be concluded with a reasonable degree of certainty that the relationship between fluorescence and membrane potential is linear in the range tested.

The specificity of the permeability and the thightness of the lipid vesicles can be tested by diluting KCl containing vesicles into isotonic sucrose solution. While there is a normal decrease in fluorescence upon increasing the outside potassium concentration, adding NaCl on the outside, to create a similar but reversed gradient for Na^+ as for K^+ does not induce a detectable change in dye fluorescence. This indicates that the vesicles are tight and that the K^+-permeability induced by valinomycin is much larger than the Na^+- and Cl^- -permeability (see Fig. 3). This finding is in agreement with Mueller and Rudin (1967), based on direct measurements of membrane potential across planar lipid bilayer containing valinomycin.

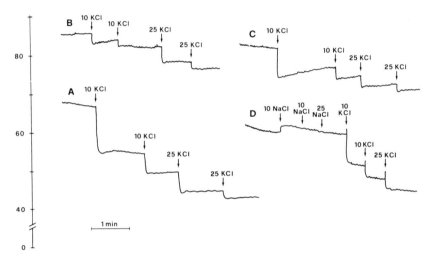

Fig. 3. Influence of acetylcholine receptor protein and α-bungarotoxin on potassium-induced fluorescence changes of phospholipid vesicles containing valinomycin. For *traces A−C* 100 μl of a vesicle suspension containing 200 mM KCl, 10 M Tris-HCl were diluted into 3 ml of 200 mM NaCl Tris HCl solution. At *arrows* X μl of 3 M KCl solution were added. In *trace A* phospholipid vesicles without acetylcholine receptor were used as a control. In *trace B* the vesicles contained the acetylcholine receptor beside valinomycin. In *trace C* 100 μl of α-bungarotoxin solution was added to the same vesicles as in trace B. For *trace D* the same vesicles used for A were diluted into 12.5% w/v sucrose solutions in order to be able to produce substantial changes in the ratio of Na_i/Na_o. The small increase in fluorescence upon adding the first 10 μl of 3 M NaCl is also seen in the absence of vesicles, it is probably due to a direct influence of a change in ionic strength to the dye fluorescence. − The experiment shows that adding 3 M NaCl, which creates a great change in the gradient for Na^+ across the vesicular membrane from $\sim 0/\sim 0$ up to $\sim 0/\sim 67$, does not affect the level of fluorescence. This finding illustrates the high degree of specificity of valinomycin-induced permeability for K^+ ions. All traces were drawn from a common zero line

Addition of Triton X-100 to a final concentration of 0.5% which disrupts all lipid vesicular structures, or omitting valinomycin, abolishes all changes in fluorescence upon changing the K^+-concentration in the suspension medium. These findings strongly suggest that the fluorescence changes measured as described above are due to changes in vesicular membrane potential.

C. Influence of a Reconstituted Protein on the Fluorescence Response Induced by a Change in the Potassium Gradient

The use of optical probes to study the influence of other ion selective ionophores or pump proteins on the membrane potential of phospholipid vesicles has been described by several authors. For review see Waggoner (1979).

We propose to use optical probes to compare the permeability status of valinomycin-containing vesicles with the permeability status of vesicles which contain a membrane protein in addition to the valinomycin, such as the acetylcholine receptor, isolated from the electric organ of electroplax.

Under the assumption that the isolated receptor protein has maintained its physiological function one would expect to be able to monitor changes in conductance of the ionophoric component of the acetylcholine receptor protein upon binding of agonists, such as carbachol or acetylcholine. One would expect to see a depolarization (decrease in fluorescence) caused by an unspecific increase of the permeability for small cations.

The experimental outcome is such that vesicles containing the acetylcholine receptor protein and valinomycin respond with a much smaller relative change in dye fluorescence upon a standard increase in outside potassium. Addition of cholinergic agonists does not produce changes in fluorescence and the size of the fluorescence changes produced by adding K^+ to the suspension medium is not dependent on the presence of agonists.

However, when the suspension medium contains α-bungarotoxin, a specific and irreversible blocker of the acetylcholine binding site, the fluorescence response is increased with respect to the response from vesicles with unblocked acetylcholine receptors (see Fig. 3).

This finding is strongly suggestive that the decreased response is not simply due to an unspecific increase in vesicle permeability due to protein insertion. Insertion of any protein is known to augment per se the permeability of lipid bilayers for ions. It is in agreement with a suggestion of Katz and Miledi (1978), that bungarotoxin might be blocking the ionophoric pore of the acetylcholine receptor protein in addition to blocking the acetylcholine binding site.

V. Comments

Figure 2 shows that changes in membrane potential of protein-free phospholipid vesicles containing valinomycin can be measured optically. The values deduced from the relative fluorescence intensity and the Nernst equation agree with those measured electrically in planar lipid bilayers by Mueller and Rudin (1967).

The fluorescence changes obtained from vesicles containing the acetylcholine receptor protein beside valinomycin upon changing the outside potassium concentration are best explained by assuming an additional unspecific cationic conductance in the range of 20%–40% of the valinomycin-mediated K^+-conductance. This additional conductance is insensitive to cholinergic agonists but it is blocked partially by α-bungarotoxin, see Fig. 3. This finding is in agreement with a suggestion of Katz and Miledi (1978) that α-bungarotoxin might be blocking the ionophoric pore beside blocking the acetylcholine binding site.

A hypothetical explanation for these unexpected properties induced by the insertion of the acetylcholine receptor might be given as follows:

The acetylcholine receptor protein, purified as described on p. 70, contains an ionophoric pore, permeable for potassium and sodium, which is in the open state.

This state might be due to the purification procedure, i.e., the use of Triton X-100 and or the affinity chromatography on the α-cobratoxin affinity column. Theoretically one has also to consider the possibility that the indodicarbocyanine dye itself might be acting as an agonist. This seems unlikely since experiments on isolated single muscle fibers did not give indication of muscle activity induced by indodicarbocyanine, as one would expect for a cholinergic agonist (Oetliker et al. 1975). Neither is there spectral indication for an interaction of the dye with the acetylcholine receptor.

Adding 20 μl valinomycin (0.5 mg/ml ethanol abs) to the motor endplate region of a frog skeletal muscle did not produce twitches while the muscle was still excitable by stimulation of the motor nerve (Oetliker, unpubl. result).

It is therefore unlikely that valinomycin exhibits agonist-like activity in the preparation used for the experiments described here.

The potentiometric method is capable of revealing changes in permeability of lipid vesicles induced by the insertion of a membrane protein. Whether it will be helpful in monitoring fast changes in permeability upon specific binding of agonists to the reconstituted acetylcholine receptor one has to awaite the outcome of future experiments with receptor preparations purified and/or reconstituted by different methods, which might preserve more of the natural function of the acetylcholine receptor.

Acknowledgments. This research was carried out under grants from Muscular Dystrophy Association and Swiss National Science Foundation Nr. 3.423-078 to H. Oetliker and Swiss National Science Foundation grant Nr. 3.032-.76 to U. Brodbeck

References

Cohen BC, Salzberg MB (1978) Rev Physiol Biochem Pharmacol 83:35–88
Hodgkin AL, Katz B (1949) J Physiol 108:37–77
Hoffmann JF, Laris PC (1974) J Physiol 239:519–552
Huang C, Mason JT (1978) Proc Natl Acad Sci USA 75:308–310
Katz B, Miledi R (1978) Proc R Soc London Ser B 203:119–133
Mueller P, Rudin DO (1967) Biochim Biophys Res Commun 26:398–404
Oetliker H (1980) J Physiol 305:26P–27P
Oetliker H, Baylor SM, Chandler WK (1975) Nature (London) 257:693–696
Russel JT, Beeler T, Martonosi A (1979) J Biol Chem 254:2040–2046
Sims PJ, Waggoner AS, Wang ChH, Hoffmann JF (1974) Biochemistry 13:3315–3330
Waggoner AS (1979) Annu Rev Biophys Bioeng 8:47–68

CD Measurements to Probe the Structure of the Purple Membrane

P.R. ALLEGRINI and P. ZAHLER

I. Introduction

A. Circular Dichroism

Plane-polarized light may be resolved into two circularly polarized components. A schematic representation of the plane-polarized wave and its relation to the circularly polarized components is shown in Fig. 1. OL and OR are the rotating components of the left- and right-hand circularly polarized light, and OA is the vector sum which represents the amplitude of the plane-polarized light. As OL and OR rotate, OA describes a sine wave motion along the axis of the light path.

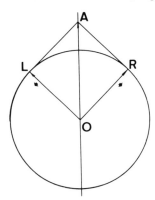

Fig. 1. Resolution of electric vector of plane-polarized light into electric vectors of right- and left-hand circularly polarized light

The absorption of circularly polarized light by optically active materials only occurs in the regions of electronic transitions. These are the regions of the Cotton effect. Then the refractive indices (n_1, n_r) of the right- and left-hand circularly polarized light are different. As a result the plane of the linear polarized light is turned by the angle α (optical rotation; Fig. 2), which is related to the refractive indices of the medium by the Fresnel equation:

$$\alpha \text{ (radians)} = \pi l \, (n_1 - n_r)/\lambda$$

(l is the path length of the light in the medium)

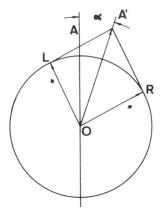

Fig. 2. When $n_l > n_r$ the left-hand circularly polarized component is slower than the right-hand component. As a result the plane of the linear polarized light is turned by the angle α

In the region of the Cotton effect the molar extinction coefficients ϵ_l and ϵ_r of the left- and righ-hand circularly polarized light are different. As a result the emergent light will be elliptically polarized. A diagrammatic representation is this situation is shown in Fig. 3 where the left- and right-hand components precess and the resultant is vector OM, which represents the intensity of transmitted light. OL' and OR' precess at different rates ($n_r \neq n_l$), hence their vector sum, OM, traces an elliptical path. The ellipticity per unit length of sample, Θ, is then defined in terms of minor and major axes of this ellipse: $\tan \Theta = OB'/OA'$. The molar ellipticity is given by the following equation:

$$[\Theta] = 3300 \, (\epsilon_l - \epsilon_r)$$

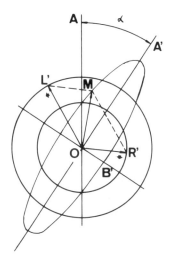

Fig. 3. Action of optically active medium on plane-polarized light. Right circularly polarized light absorbed to a greater extent than left, and index of refraction for left greater than for right, as depicted

Θ can be positive or negative in sign depending upon the relative magnitude of the molar extinction coefficients ϵ_l and ϵ_r.

The advantage of the CD technique over ORD (optical rotary dispersion) is that these Cotton effects may be more clearly resolved and identified with the underlying electronic transition because of the "sharpness" associated with the absorption phenomena compared with dispersive phenomena.

For further details see: A.G.Walton and J.Blackwell (1973) Biopolymers (Academic Press, London New York), pp 262

B. Bacteriorhodopsin

Bacteriorhodopsin, the only polypeptide in the purple membrane of *Halobacterium halobium,* acts as a light-driven proton pump. For recent reviews see Henderson (1977) and Stoeckenius et al. (1979).

Bacteriorhodopsin molecules are arranged in a two-dimensional hexagonal lattice (Henderson 1975) in distinct patches in its cell membrane. The three-dimensional structure has been obtained at a resolution of 7 Å (Henderson and Unwin 1975). Rotational diffusion measurements show that bacteriorhodopsin is immobilized in the purple membrane (Cherry et al. 1977), as expected from the existence of the crystalline lattice.

Characteristic exciton bands occur in the visible CD spectrum of purple membrane suspensions. They are due to the transition charge interactions between the retinal chromophores of adjacent bacteriorhodopsin molecules in the hexagonal lattice of the purple membrane (Cherry et al. 1978).

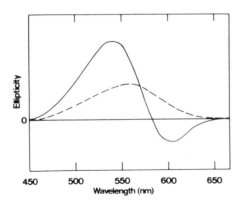

Fig. 4. CD spectrum of bacteriorhodopsin in the visible range. (————) bacteriorhodopsin in the crystalline lattice; (– – –) monomeric bacteriorhodopsin (Cherry et al. 1978)

Figure 4 shows the CD spectrum of purple membrane suspensions in the visible range. The two bands of opposite sign with crossover at 574 nm are the result of the transition of a broad positive band centered at the chromophore absorption maximum of bacteriorhodopsin (567 nm) which is due to the interaction of the retinyl residue with the protein part in the monomeric form to an exciton band with a positive short wavelength lobe and a negative long wavelength lobe, which is due to the interaction between retinal chromophores on adjacent bacteriorhodopsin molecules in the crystalline lattice of the purple membrane.

These exciton effects are not observed when bacteriorhodopsin is solubilized into micelles containing protein monomers or when each bacteriorhodopsin in the protein lattice is surrounded by bacterio-opsin molecules only (by lack of retinal) (Heyn et al. 1975). It can be shown that independent rotational motion of each bacteriorhodopsin molecule about an axis perpendicular to the plane of the membrane is sufficient to cause the exciton effect to disappear. Even if the average chromophore-chromophore distance is small enough for an effective interaction, rotational motion or rotational disorder leads to a complete averaging out of the angular dependence of the exciton amplitudes. The exciton effect may thus be used to distinguish crystalline or specifically aggregated bacteriorhodopsin from monomeric bacteriorhodopsin.

II. Experiments

A. Materials

basal salt solution: 250 g NaCl/l; 20 g MgSO$_4$ · 7 H$_2$O/l; 3 g trisodium citrate · 2 H$_2$O/l; 2 g KCl/l
0.1 M NaCl
30% sucrose; 60% sucrose
phosphate buffer: 500 mM pH 7.0
20% Na$_2$CO$_3$ in 1 N NaOH
DNase (bovine pancreas, Sigma)
Triton X-100 (Merck)
Dipalmitoyllecithin (Fluka)
Sephadex G-200 column (diameter 2 cm; length 40 cm)
tissue grinder (tight-fitting Teflon pestle; Thomas, Philadelphia, USA)
water bath type sonifier (Laboratory Supplies Company, Hicksville, N.Y.)
Roussel-Jouan Dichrograph II
spectrophotometer
Sorvall centrifuge (Superspeed RC2-B) with SS-34 rotor
an ultracentrifuge with a SW 27 rotor

B. Isolation of the Purple Membrane

The isolation of bacteriorhodopsin is made according to D. Oesterhelt and W. Stoeckenius (1974):

Cells from a 10 liter culture are harvested by centrifugation (15 min, 13,000 g) and resuspended in 250 ml basal salt solution. The cells are then dialyzed with 5 mg DNase against 2 l of 0.1 M NaCl. The lysate is centrifuged at 40,000 g for 40 min and the red supernatant decanted. The reddish purple sediment is resuspended twice in 300 ml of 0.1 M NaCl and twice in deionized water using a tight-fitting Teflon pestle and each time centrifuged under the same conditions as above. The final sediment, resuspended in 6–10 ml of water, is layered over a linear 30%–50% sucrose density gradient and centrifuged at 100,000 g for 17 h.

The purple band (1.18 g/cm^3) is collected and sucrose is removed by dilution to 300 ml and repeated centrifugation at 50,000 g for 30 min.

C. Solubilization of Bacteriorhodopsin with Triton X-100

A purple membrane suspension is mixed with phosphate buffer pH 7.0 to a final concentration of 6–7 mg protein/ml and 25 mM phosphate buffer. This suspension is sonified in a water bath type sonifier for 1 min. A CD spectrum is carried out with a Roussel-Jouan Dichrograph II spectropolarimeter at room temperature in a 0.1 cm cell from Hellma.

To the purple membrane suspension 5%–10% (v:v) Triton X-100 is added. After an incubation time of about 24 h at room temperature the bacteriorhodopsin is solubilized in a monomeric form. A CD spectrum is made of this solution in the same manner as above.

D. Temperature-Dependent Aggregation of Bacteriorhodopsin in Dipalmitoyllecithin Vesicles

To the solubilized purple membranes a twofold ratio by weight of dipalmitoyllecithin is added and the whole is agitated for a few minutes to solubilize the lipid. To remove the Triton X-100 the mixture is placed on a Sephadex G-200 column and the vesicles are eluted with 25 mM phosphate buffer pH 7.0. The Triton content in the fractions is determined by adding 0.5 ml of 20% Na$_2$CO$_3$ in 1 N NaOH to 1 ml of each fraction (Rottem et al. 1968). Turbidity indicates the presence of Triton. A typical elution diagram is shown in Fig. 5.

The pooled fractions with the bacteriorhodopsin in dipalmitoyllecithin vesicles are evaporated to a protein content of 2 mg protein/ml.

Visible CD spectra of these vesicles are made above and below the lipid phase transition ($\sim 0°C; \sim 45°C$) in a 1-cm cell (see also Cherry et al. 1978).

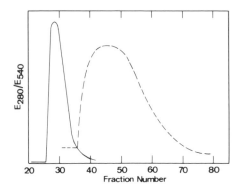

Fig. 5. Elution diagram of the Sepha-dex G-200 column. (———) protein content (E_{280}); (– – –) Triton X-100 content (E_{540})

Compare the peak-to-peak (535 nm, 605 nm) molar ellipticity $[\Theta]$ of the purple membranes and the bacteriorhodopsin in dipalmitoyllecithin vesicles:

$$[\Theta] = 3300 \times \Delta\epsilon \ [deg \times cm^2/dmol]$$

whereas $\Delta\epsilon$ can be calculated by the following equation:

$$\Delta\epsilon = \frac{d \times s \times M}{c \times 1}$$

Abbreviations: d: pen deflection in mm; s: sensibility; l: pathlength of the cell in cm; M: molecular weight of bacteriorhodopsin (26,534 daltons) (cor-responding to 1 mol of retinal); c: concentration in mg/ml

Sample	Wavelength (nm)	Pathlength (cm)	Sensibility	Conc. (mg/ml)	Pen deflection (mm)	$[\Theta]$ (deg x cm^2/dmol)

Sample	Peak to peak molar ellipticity $\Theta_{535} - \Theta_{605}$

References

Cherry RJ, Heyn MP, Oesterhelt D (1977) Rotational diffusion and exciton coupling of bacteriorhodopsin in the cell membrane of *Halobacterium halobium*. FEBS Lett 78:25–30

Cherry RJ, Müller U, Henderson R, Heyn MP (1978) Temperature dependent aggregation of bacteriorhodopsin in dipalmitoyl- and dimyristoylphosphatidylcholine vesicles. J Mol Biol 121:283–299

Henderson R (1975) The structure of the purple membrane from *Halobacterium halobium:* Analysis of the X-ray diffraction pattern. J Mol Biol 93:123–138

Henderson R (1977) The purple membrane from *Halobacterium halobium*. Annu Rev Biophys Bioeng 6:87–109

Henderson R, Unwin PNT (1975) Three-dimensional model of purple membrane obtained by electron microscopy. Nature (London) 257:28–32

Heyn MP, Bauer P-J, Dencher NA (1975) A natural CD label to probe the structure of the purple membrane from *Halobacterium halobium* by means of exciton coupling effects. Biochem Biophys Res Commun 67:897–903

Oesterhelt D, Stoeckenius W (1974) Isolation of the cell membrane of *Halobacterium halobium* and its fractionation into red and purple membrane. Methods in Enzymology XXXI: 667–678

Rottem S, Stein O, Razin S (1968) Reassembly of *Mycoplasma* membranes disaggregated by detergents. Arch Biochem Biophys 125:46–56

Stoeckenius W, Lozier RH, Bogomolni RA (1979) Bacteriorhodopsin and the purple membrane of halobacteria. Biochim Biophys Acta 505:215–278

Characterization – Other Techniques

Lipid-Protein Interactions in Monolayers at the Air-Water Interface

F. PATTUS and C. ROTHEN

I. Introduction and Aims

Among a wide variety of techniques used for studying lipid protein interactions and biological activities at the lipid-water interface, the monolayer techniques is extremely interesting, With this system, it is possible to vary and measure the lipid packing without changing the composition of the interface.

As a membrane model, a phospholipid monolayer may be considered as half a bilayer. The orientation of the phospholipid molecules are quite similar to those in biological membranes.

A phospholipid monolayer is characterized by its isotherm of compression (surface pressure versus area occupied by one phospholipid molecule). Depending on the chemical structure of the phospholipid and temperature phospholipid monolayers can be in a lqiuid expanded or solid condensed state. Phase transitions between these two states may occur. For review see Davies and Rideal (1961), Pethica (1968), Jackson (1970).

Two types of experimental approaches will be presented.

Expt. I: Interaction of a soluble protein with lipid monolayers.

Expt. II: Formation of lipoprotein films at the air-water interface from biological membranes or reconstituted membrane complexes. Influence of lipid packing on a membrane-bound enzymatic activity.

II. Equipment and Precautions

Measurement of Surface Pressure

Surface pressure by measuring a downward force (F) on a platinum plate crossing the interface (Fig. 1) connected to an electrobalance (or a transducer) (see Fig. 2).

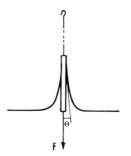

Fig. 1. Measurement of surface pressure by the Wilhelmy-plate-method

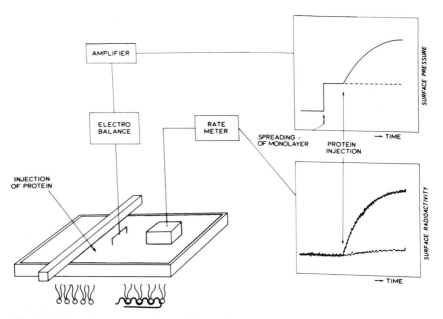

Fig. 2. Schematic representation of the technique used to study the interaction of (radioactive) proteins with a monomolecular layer of lipids

$F = 2\,l\,\gamma \cos \theta$ where γ = interfacial tension in dynes/cm

$\qquad\qquad\qquad\quad \theta$ = contact angle

$\qquad\qquad\qquad\quad l$ = length of the Pt plate

If the Pt plate is perfectly cleaned $\theta = 0$, $\cos \theta = 1$, $l = 1.962$ cm for a Prolabo (France) Pt plate (Catalog No. 03294.00).

If m is the mass measured by the balance

$$\gamma = \frac{mg}{2l} = \frac{m}{2}\,\frac{981}{1.962} = \frac{m\,10^3}{4} \qquad \text{1 dyne/cm corresponds to 4 mg}$$

For a pure air/water interface (without a monolayer)

$$\gamma = \gamma_0 = 72.8 \text{ dynes/cm at } 25°C$$

In presence of a lipid monolayer

$$\gamma = \gamma_m \quad \text{where } \gamma_m < \gamma_0$$

Surface pressure is defined as

$$\pi = \gamma_0 - \gamma_m \quad \text{when } \gamma = \gamma_0 \quad \pi = 0$$

Different types of electrobalance with an output for a recorder can be used (Beckman LM 600, Mettler PC 180.03) or transducers (Stoutham Universal Transducing Cell, Model UC-2) connected to appropriate accessories.

Monolayer Trough

For Experiment I a simple rectangular Teflon trough (10–20 ml volume, 1 cm depth) with a Teflon magnetic bar for stirring can be used. The surface of the trough edges must be perfectly polished in order to be able to clean the surface with a Teflon barrier.

For Experiment II a four-compartment trough is required (Rietsch et al. 1977) (Fig. 3).

The total monolayer apparatus has been described (Verger and de Haas 1973). It consists of a thermostated box, a device to move a mobile barrier in order to compress or expand the monolayer on electrobalance to measure surface pressure and a regulatory unit to maintain surface pressure constant by moving the barrier.

Precautions. The monolayer technique is very sensitive to tensioactive impurities. The water used must be bidistillated in potassium permanganate. Buffers must be made with high quality substances.

The spreading solutions of phospholipids (in chloroform, chloroform methanol of hexane) must be done in high purity dried solvents to avoid spontaneous hydrolysis and kept under nitrogen atmosphere.

Since protein-lipid interactions are very sensitive to small changes in the properties of the film, the phospholipid used must be pure. Most commercial phospholipids contain fatty acids and lyso compounds. They must be purified further on silicic acid columns or preparative thin layer chromatography.

Phospholipids used were from Serdary Research Lab. Inc. London, Ontario, Canada, and from Medmark Res. Chem Grünwald, Munich, W. Germany.

When unsaturated phospholipids are used it is necessary to work under argon or nitrogen atmosphere.

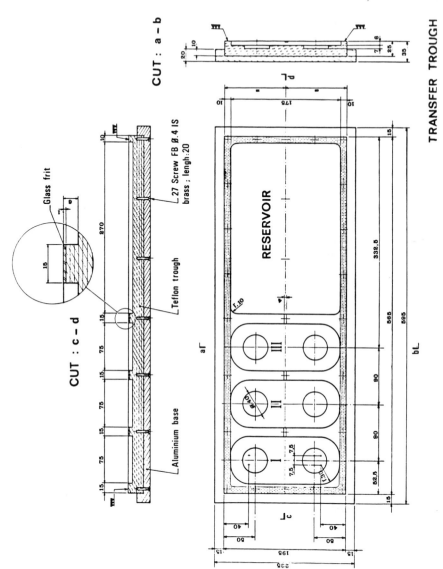

Fig. 3. Special Teflon trough used for spreading membranes and for film transfer

III. Experimental Procedures

A. Expt. I. Interaction of a Cardiotoxin from Naja Mossambica Venom with Phospholipid Monolayers

Snake venom toxins can be separated according to their pharmacological effects into neuro and cardio toxins. Both types of toxin have potent effects on membranes, but the neurotoxins are highly specific for only a few cell targets, whereas cardiotoxins display a general action on many kinds of membranes (Yang 1974). Cardiotoxins are described as direct lytic factors, due to their ability to break down the plasmic membrane. Lipids have been implicated in the binding of cardiotoxin to cell membranes.

The purpose of the monolayer experiments is to show that Cardiotoxin II from this venom has a high penetration capacity in phospholipid monolayers and shows a net specificity for negatively charged phospholipids (P. Bougis, pers. comm.).

1. Principle (see Fig. 2 taken from R.A. Demel et al. 1974)

The water-soluble protein is injected under a monomolecular film of phospholipid at a given pressure (π_0). The penetration of the protein at the interface is followed as a function of time by measuring the increase of surface pressure. Two parameters give information about the penetration:

1. the equilibrium surface pressure π_e, expressed as the increase of surface pressure $\Delta \pi = \pi_e - \pi_0$
2. the initial velocity of the pressure increase $\left(\dfrac{d\pi}{dt} \right)_{t = 0}$. The penetration capacity of the protein is given by the lipid packing (given by the surface pressure-lipid packing relationship of the phospholipid monolayer:isotherm of compression) where the toxin can no longer penetrate.

2. Experimental Procedure

Fill a 10 ml Teflon trough with 1 mM Tris HCl, 10 mM NaCl, 1 mM CaCl$_2$, pH 7.5 buffer. Clean the surface by suction of water at the surface with a Pasteur pipette connected to water pump while moving the barrier over the entire plane of the trough. Remove the platinum plate from the chromosulfuric acid mixture, rinse it with bidistilled water, and attach it to the electro balance, making sure that the platinum plate crosses the air-water interface. Adjust the recorder pen to zero surface pressure (pure air-water interface).

Spread the phospholipid with a syringe by slow deposition of the chloroform solution on the interface until a surface pressure $\pi_0 = 25$ dynes/cm, for example, is obtained. Switch on the magnetic stirrer. 50 μg of toxin is then injected into the subphase and surface pressure is recorded until equilibrium. After each experiment the platinum plate is placed in chromosulfuric acid. The trough is then washed with distilled water, ethanol, and finally with bidistilled water.

The following lipids can be used:

phosphatidyl glycerol
total lipid extract from *E. coli*
egg lecithin
phosphatidyl ethanol amine
cholesterol

Plot $\Delta\pi$ and $\left(\dfrac{d\pi}{dt}\right)_{t=0}$ as function of surface pressure (π_0).

Lipid used π_0 π_e $\Delta\pi$ $\left(\dfrac{d\pi}{dt}\right)_0$

B. Expt. II. Spreading of Membranes of the Air-Water Interface

Most of the intrinsic proteins are insoluble in water and detergents are needed to disperse them. The detergents interfere with surface pressure measurements and thus the type of experiment described in Expt. I is not possible. If detergents are avoided, the proteins may aggregate without any affinity for lipid. The following experiments show how one can obtain lipoprotein films at the air-water interface. These films are derived from biological membranes or reconstituted lipoprotein complexes.

For further information see Verger and Pattus (1976), Pattus et al. (1978).

1. Principle

A suspension of membrane vesicles or reconstituted lipoprotein is added slowly, dropwise, along a clean wet glass rod crossing the interface. (Fig. 4). The membrane suspension should drip from the rod and spread over the water surface. When the water surface is large enough, no rise in surface pressure is observed. The glass rod is then removed and the mobile barrier moves along to compress the surface film. The film then passes through a second compartment where it is rinsed and when further compressed passes completely to the surface of the third compartment. At this stage the film can be recovered and analyzed for its protein content or for enzymatic activities, or a substrate or an effector (proteolytic enzymes, ions, inhibitors) may be injected into the subphase below the film.

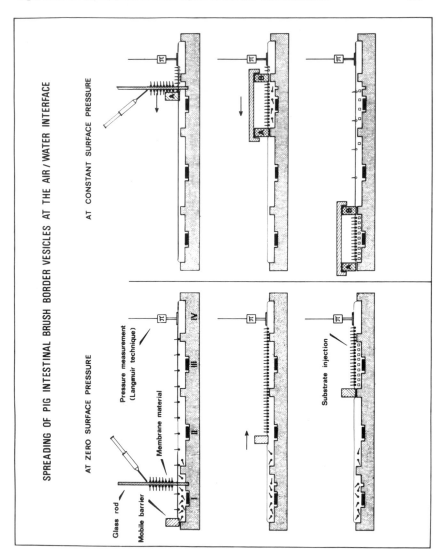

Fig. 4. Principle of two different spreading techniques used. The thermostated (25°C) Teflon trough is composed of four compartments connected by narrow surface canals. The subphase of compartments I, II and III are mixed by magnetic stirrers at 250 rpm. Compartment II is used as a rinsing compartment. The surface pressure recording device is based on the principle of the Longmuir balance. This device is located in compartment IV

In the following experiments human red cell membranes will be used as starting membrane material. The compression isotherm of the films obtained will be measured. The acetylcholine esterase activity will be measured after recovery of the film or by injection of the substrate (acetylthiocholine) below the film. The film will also be deposited on a mica plate and visualized after freeze etching by electron microscopy (see Sect. IV.A).

2. Experimental Procedure

Fill the 4th compartment trough with 10 mM Tris HCl, 0.1 m NaCl, 10 mM $CaCl_2$ buffer, pH 7.4. Clean the surface by moving the barrier along the trough while removing the water at the surface by aspiration with a Pasteur pipette connected to a vacuum. Treat the platinum plate as described in Expt. I. Take the glass rod from the chromolsulfuric acid solution and wash it extensively with tap water and bidistilled water. Place it in compartment I. Adjust the pen (right) corresponding to surface pressure, to zero. 10 μl of the ghost suspension (10 mg/ml of protein) is then added onto the glass rod, 2 cm from the interface. After a spreading time of 1 min (or 25 min) the film is slowly compressed. Surface pressure and advancement of the barrier is recorded when the film is over compartment IV (rectangular compartment).

a) Recovery of the Film Acetylcholine Esterase Activity. The barrier is stopped when π = 30.0 dynes/cm. Weigh the trap (M_o). Connect it to a vacuum and recover the film by aspiration while compressing it with the barrier (Fig. 5). Weigh the trap (M). $M - M_o$ = "volume" of buffer containing the film (V). Take the same volume of subphase. Fill the UV cell with 1 ml of the film solution and place it in the UV spectrophotometer. Add 10 μl of DTNB (5 mg/ml) and 10 μl acetylthiocholine (30 mg/ml). Record the kinetics at 412 nm. $\epsilon_{412 \text{ nm}}$ = 13,600.

$$\text{Acetylcholine esterase units in V ml} = \frac{(\text{OD/min}) \times (\text{V in ml})}{13.6}$$

Do the same with 1 ml of the subphase (blank) and with 10 μl of the ghost suspension (diluted 10 times) in 1 ml buffer. The spreading yield of acetylcholine esterase activity is:

$$\text{Spreading yield } \% = \frac{\text{units recovered from the surface}}{\text{units in 10 } \mu\text{l ghost}} \times 100$$

$$\text{Surface density of acetylcholine esterase activity} = \frac{\text{units recovered}}{\text{surface of the film}}$$

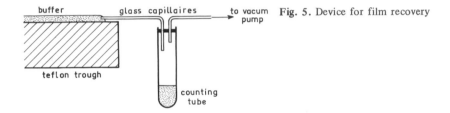

Fig. 5. Device for film recovery

Plot the compression isotherm of the film.

π dynes/cm	L (cm) advancement of the barrier	S (cm^2) surface of the film	unit/cm^2

Total surface of compartments II, III and IV = 210 cm^2. Size of compartment I is $17.5 \times L$ (cm)

b) Acetylcholine Esterase Activity in Situ. In this experiment the glass rod is located on compartment IV and the film compressed to compartment I which is stirred. The barrier ist stopped when $\pi = 30{,}0$ dynes/cm.

1 ml of DTNB (5 mg/l) is injected in the subphase of compartment IV. Then at T = 0, 1 ml of acetylthiocholine (30 mg/ml) is injected in the subphase. At different times take 1 ml of the subphase and read the O.D at 412 nm. After 1.5 h remove the film from the surface and measure the blank.

Plot O.D as a function of time and calculate the surface density of acetylcholine esterase.

Volume of compartment IV = 100 ml
Surface of compartment IV = 210 cm^2

$$\text{Units/cm}^2 = \frac{(\text{O.D/min}) \times 110}{13.6 \times 210}$$

Compare to the surface density measured after recovery.

C. Dipping Technique: Film Deposition on Mica Plates

1. Principle (see Fig. 6a)

Blodgett (1935) describes in her paper a method for depositing stearate monolayer films on glass plates. In our example two lipoprotein monolayers are deposited on a mica plate.

The hydrophilic mica is first dipped into the subphase exposing the hydrophilic site of the film. It is then possible to deposit a monolayer on the mica plate.

In the last step the "hydrophobic" mica plate is again dipped into the subphase, thereby forming a bilayer. This is then representative of a model membrane.

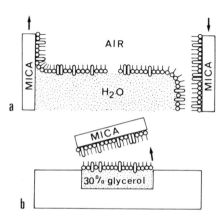

Fig. 6. a Dipping technique with the hydrophilic mica using a micromanipulator. b Fracturing of the bilayer by tearing off the mica from the U-shaped brass holder. The second monolayer remains on the frozen 30% glycerol

2. Experimental Procedure

Place a freshly split mica plate on a pair of microtweezers and dip it through the film (see Sect. 2.2) into the subphase with a micromanipulator. One should not observe any changes in surface pressure during the procedure. Draw the mica plate slowly up through the film. This should deposit the first layer with te lipid polar group oriented toward the mica.

Air-dry the plate for several minutes and then dip it back very quickly through the monolayer, thus depositing the second layer.

After each "film-dipping" one should notice a decrease in surface pressure. Compensate the initial surface pressure with a movable barrier.

Dipping	$\Delta\pi$ (dynes/cm)
1 air-water	
2 water-air	
3 air-water	

D. Bilayer Fracturing and Visualization of the Hydrophobic Site of the Model Membrane

This section will be demonstrated (Zahler et al. 1976).

1. A Modified Freeze Fracturing and Freeze Etching Method

In the last 15 years the freeze etching process has developed into a general method for the preparation of biological material for examination by electron microscopy. It is composed of 4 steps:

freezing
fracturing
etching
coating

By rapid freezing the molecular structure of biological mebranes does not change.

The fracturing of membranes is of special interest because the membranes will split preferentially in the hydrophobic region.

The etching is due to the membranes (lipids, proteins, carbohydrates) as a consequence of the high vacuum and the large temperature gradient between the sample and the knife assembly.

The evaporation of platinum at a certain angle on the biological membrane will then show the prominent three-dimensional structure of the biological material.

The evaporation with carbon results in an exact replica of the treated material.

Deamer and Branton (1967) demonstrated on stearate multi-monolayers that they will also fracture at low pressure and low temperature, preferentially in the hydrophobic region.

In our experiment the bilayer is split into two monolayers according to this principle.

A special fracturing device will be used for this experiment (see Fig. 7).

The mica plate with the deposited bilayer will transfer in water to a large container where it will attach itself to the U-shaped fracturing device. The mica holder is rinsed with 30% glycerol, rapidly frozen in Freon 21 and then placed on the pre-cooled ($-150°C$) stage of a Balzers BA 360 freeze-etch apparatus. Under vacuum (10^{-6} Torr) the mica plate will break away from the frozen glycerol with the knife assembly.

The exposed surface of a hydrophobic membrane site (see Fig. 6b) is etched briefly (60 s at $-100°C$) and shadowed with platinum and carbon.

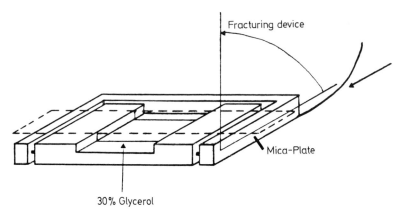

Fig. 7. Device for fracturing the frozen bilayer

2. Electron Microscopy

The carbon replica, corresponding to the hydrophobic fracture face of the second monolayer deposited on the mica (see Fig. 6b) is placed on a 200-mesh copper net and used for electron microscopy (Phillips EM 200 at 60 kV).

The carbon replica has a resolving power in the range of 2 Å, therefore it is possible to observe lipoprotein complexes as particles (diameter 80 Å) on the hydrophic site of the membrane.

The preparations are examined under the electron microscope with a magnification of ca. 30,000 X.

References

Blodgett K (1935) J Am Chem Soc 157:1007

Davies D, Rideal EK (1961) Interfacial phenomena. Academic Press, London New York

Deamer DN, Branton D (1967) Science 158:655

Demel RA, London Y, Geurts van Kessel, van Deenen LLM (1974) In: Bolis L, Bloch K, Lurier SE, Lyner F (eds) Comparative biochemistry and physiology of transport. North Holland Publ Co, Amsterdam, pp 25–32

Jackson DM (1970) In: Livia Bolis (ed) Permeability and function of biological membranes. North Holland Publ Co, Amsterdam London

Pattus F, Desnuelle P, Verger R (1978) Biochim Biophys Acta 507:62–82

Pethica BA (1968) In: Tria E, Scanu A (eds) Structural and functional aspects of lipoproteins in living systems. Academic Press, London New York, pp 37–72

Rietsch J, Pattus F, Desnuelle P, Verger R (1977) J Biol Chem 252:4313–4318

Verger R, de Haas GH (1973) Chem Phys Lipids 10:127–136

Verger R, Pattus F (1976) Chem Phys Lipids 16:285–291

Yang CC (1974) Toxicon 12:1

Zahler P, Rothen C, Flückiger R (1976) Chimia 30:85

Incorporation of the Polypeptide Hormone ACTH into Planar Lipid Bilayers: Evaluation by the Capacitance Minimization Technique

P. SCHOCH and D. SARGENT

I. Introduction

Polypeptide-lipid interactions are of great importance for many processes controlling the state of biological membranes. Several methods, most of them spectroscopic, using liposomes or micelles as model membranes, are available for studying these interactions. This experiment uses a new approach, applicable to planar lipid bilayers, which allows the determination of binding by monitoring the change of electrostatic surface potentials at the membrane-solution interface (Schoch et al. 1979).

A. Electrostatic Potentials at the Membrane-Solution Interface

Figure 1 shows the surface potentials that may arise at the membrane-solution interface. In the case of fixed charges — either from charged lipid head-groups or from bound molecules — there will be a so-called Gouy-Chapman fixed-charge surface potential, V_G. It depends on the surface charge density and on the ionic strength in the adjacent aqueous solution.

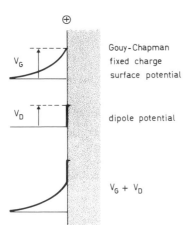

Fig. 1. Electrostatic potentials at the membrane-solution interface

Generally there will also be a dipole potential jump, V_D, caused by the ordered arrangement of the molecules involved. V_D depends on the dipole moment per molecule, the number of dipoles per unit area and the local dielectric constant, but is independent of the ionic strength in the solution. For a detailed description see Aveyard and Haydon (1973) and McLaughlin (1977).

The formulas describing these surface potentials, adapted to convenient units, are:

$$V_G = 50.8 \cdot \ln(s + \sqrt{s^2 + 1})$$

$$s = 136 \cdot \sigma / \sqrt{c}$$

V_G: fixed-charge surface potential, in mV

σ: surface charge density, in elementary charges per Å^2

c: concentration of 1-1-electrolyte, in M/l

or, solving for surface charge density

$$\sigma = \sqrt{c}/136 \cdot \sinh(V_G/50.8)$$

$$V_D = 3.77 \cdot 10^4 \cdot D \cdot \mu/\epsilon_a$$

V_D: dipole potential jump, in mV

D: dipole density, in dipoles/Å^2

μ: dipole moment per molecule, in Debye

ϵ_a: dielectric constant of the adsorption region

These equations are numerically valid at room temperature.

B. Membrane Capacitance

The electrical capacitance of a lipid bilayer may be compared with that of a thin parallel-plate capacitor. However, in contrast to a capacitor, bilayer capacitance is voltage-dependent because of its compressibility (variable thickness):

$$C_m(V) = \frac{\epsilon_o \, \epsilon_m \, A_m}{d_m(V)}$$

C_m: membrane capacitance

ϵ_o: permittivity of free space

ϵ_m: dielectric constant of the lipophilic core of the membrane

A_m: area of the membrane

d_m: thickness of the core of the membrane

Compressible membranes show an increase of capacitance proportional to the square of the voltage across them. As the factor determining the compression is the potential difference between the two surfaces, both the externally applied voltage and the difference of the surface potentials on the two sides of the membrane is involved. Figure 2 shows how the surface potentials influence the capacitance-voltage characteristic. In the first case the surface potentials on both sides are equal, resulting in a symmetrical potential profile. Here the membrane has its minimum capacitance at zero applied voltage. In the second case the two surface potentials are unequal, which results in an intrinsic compressive potential difference across the membrane interior when the aqueous solutions are at the same potential. The essential point is that this intrinsic electric field may be adjusted to zero by an externally applied voltage. This voltage, the "capacitance minimization potential", V_{Cmin}, reduces the membrane capacitance to its minimal value. It reflects the asymmetry in the surface potentials of the two sides of the membrane caused, for example, by binding of a membrane-active peptide. From the determination of V_{Cmin} at different ionic strengths, fixed-charge surface potentials may be distinguished from dipole potentials. Whereas the fixed-charge surface potentials can be determined individually on each side of the membrane, only the *difference* of the dipole potentials can be measured.

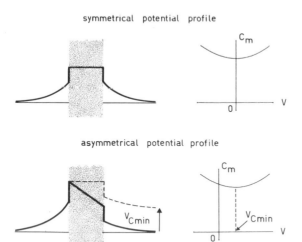

Fig. 2. Dependence of capacitance minimization potential, V_{Cmin}, on surface potentials

C. Thermodynamic Description of the Binding of Charged Molecules to Neutral Lipid Bilayers

Quite generally the interaction of a molecule A with a membrane may be represented by the following two-step equilibrium:

$$A_{free} \rightleftharpoons A_{adsorbed} \rightleftharpoons A_{incorporated}$$

Two bound states, "adsorbed" and "incorporated", are distinguished, where "incorporation" implies an interaction involving more than just a surface association. For ACTH the molecules appear to span the membrane, with parts of them appearing on the opposite side of the bilayer. Thus, while adsorbed ACTH molecules affect the surface potential only on the cis-side, incorporated ones affect it on both the cis- and trans-sides.

The equilibrium concentrations are governed by

1. the changes of the standard free energy (binding energies) $\Delta G_{o\ free-ad}$ and $\Delta G_{o\ ad-in}$, and
2. the electrostatic repulsion between neighboring bound molecules on the membrane.

The calculation of such electrical repulsion is usually done using the Gouy-Chapman theory, which assumes smeared surface charges. Although it will not be obvious in the qualitative analysis done in this experiment, it is interesting to note that the Gouy-Chapman treatment is not sufficient to explain the data: the *discrete* nature of the charge distribution must be explicitly taken into account. For a detailed treatment of the thermodynamic analysis of binding of charged species see Schoch et al. (1980, 1981).

Figure 3 represents in a general manner an example of a possible potential profile resulting from adsorption and incorporation of a charged molecule to a bilayer.

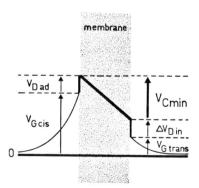

Fig. 3. Possible potential profile resulting from adsorption and incorporation of a charged molecule into a bilayer

II. Equipment and Solutions

A. Equipment

glassware

100 μl automatic pipette
10 ml graduated cylinder
Pasteur pipettes

electronic apparatus

Figure 4 shows a block diagram of the measuring apparatus. The measurement chamber consists of two halves separated by a thin Teflon septum (30 μm thick) into which a small hole of about 0.06–0.08 mm^2 was melted by a glowing filament. The Ag/AgCl reference electrodes are connected to the measurement chambers by agar-salt bridges. The salt bridges have a resistence of about 1 kΩ and are the limiting resistance in the external circuit. The device allows one to measure capacitance and current as a function of the applied voltage, as well as to continuously record the capacitance minimization potential, V_{Cmin}. Both C_m and V_{Cmin} are monitored on a chart recorder.

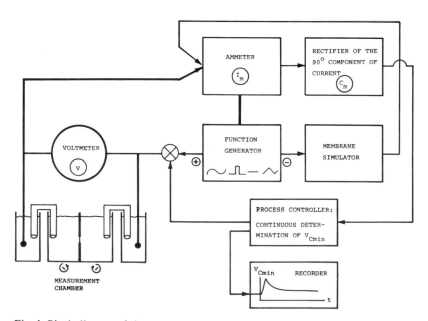

Fig. 4. Block diagram of the measuring apparatus

B. Solutions

electrolyte solutions: 9 mM KCl, 2 mM imidazole, pH 7.0 (ionic strength
                          10 mM)
                          4 M KCl
lipid solutions: 0.7% egg lecithin in hexane
$ACTH^{6+}_{1-24}$ (MW 2934):   8 mM in 10 mM KCl

III. Experimental Procedures

A. Preparation for Measurements

1. Formation of Bilayer Membranes

Lipid bilayer membranes are formed essentially by the technique of Montal
and Mueller (1972): two separately formed monolayers are apposed over an
aperture in a thin septum. The prodecure is as follows. Measure out 10 ml of
the 10 mM aqueous electrolyte solution. Pipette enough of this solution
into both half-chambers so that the levels stand 2 mm below the hole in the
septum. Add two drops of the lipid/hexane solution to the aqueous surfaces
for the formation of surface "monolayers". As we are interested in easily
compressible membranes, try to form the bilayer after about 1 min, before
all the hexane has evaporated. This is done by slowly raising the aqueous
levels to about 2 mm above the hole on both sides, one after the other.
Whether or not a membrane has been formed can be recognized by the capacitance signal on the recorder: if the membrane breaks or none is formed
the pen kicks off-scale. In this case try again by lowering the aqueous level
on one side to beneath the hole, and then raise it again. Once you have suc-
ceeded in forming a bilayer, fill in the rest of the aqueous solution so that
both chambers contain 5 ml. Any bulging of the membrane is removed by
adjusting the aqueous level so that the capacitance becomes minimal.

2. Determination of V_{Cmin}

Membrane capacitance is measured using a 1000 Hz ac signal of about
20 mV (peak-peak). Apply this signal and check it on the oscilloscope.
V_{Cmin} can be monitored and recorded by hand, but the apparatus provided
does this automatically with a process controller: the voltage is continuously
varied to determine the value that results in minimal capacitance. As the

change in capacitance is very small compared with the total capacitance, the bulk of the membrane capacitance is compensated using the membrane simulator. The simulator is supplied with a 1000 Hz signal opposite in phase to that of the membrane. As both are connected to the same ammeter the simulator capacitance signal is subtracted from that of the membrane. When a bilayer membrane has been formed adjust the membrane simulator to compensate the membrane capacitance signal down to only a few pF, then put the process controller into operation. The base-line for both V_{Cmin} (should be 0 ± 2 mV) and C_m is recorded for a few minutes before beginning an experiment.

B. Binding of $ACTH^{6+}_{1-24}$ to Lipid Bilayers

(Typical results are given in Fig. 5)

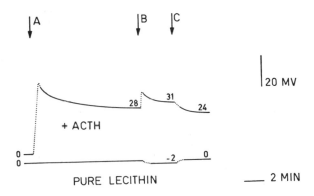

Fig. 5. Changes in capacitance minimization potential with time on addition of ACTH. Following the addition of ACTH (at time A) V_{Cmin} rapidly rises, reflecting the adsorption to the cis-side. This is followed by a much slower incorporation reaction in which part of the molecule is transferred to the trans-side (subsequent decline of V_{Cmin}). When the ionic strength on the trans-side is increased to 90 mM the trans positive charges are shielded to a greater extent, causing the second increase (time B). This change in the electrostatic energy barrier disturbs the previous equilibrium, however, and the slow incorporation reaction again results in a subsequent decline of V_{Cmin}. When the ionic strength on the cis-side is raised (C), the cis charges are shielded, resulting in a decrease of V_{Cmin}. The reduced energy barrier on the cis-side also allows more ACTH to bind: without this effect the change in V_{Cmin} after time C would be larger

1. Determination of Surface Charge of Unmodified Lipid Membranes

The surface charge density on each side of the membrane is determined by the dependence of the fixed-charge surface potential on the ionic strength. Increase the salt concentration in the front chamber from 10 to 90 mM by removing 100 μl of solution with the sampler and replacing it by the same amount of 4 M KCl. After careful stirring, V_{Cmin} is allowed to equilibrate. As this measured change in V_{Cmin} corresponds to the change of the Gouy-Chapman surface potential upon a change of the ionic strength at *constant* surface charge, the charge density may be determined by fitting the observed ΔV_{Cmin} between the two corresponding curves of Fig. 6a,b.

2. Binding of $ACTH^{6+}_{1-24}$ to Lipid Bilayer Membranes at 10 mM Electrolyte Concentration

Remove 100 μl of the front chamber solution and replace it by the same amount of 8 mM ACTH, resulting in a free ACTH concentration in solution of $1.6 \cdot 10^{-4}$ M. Stir cautiously for about 10 s and record the developing V_{Cmin} until a constant level is reached (several minutes).

3. Demonstration of a "Trans Effect" upon ACTH-Binding

Once the binding of ACTH to be bilayer has come to equilibrium increase the ionic strength on the trans-side of the membrane from 10 to 90 mM by removal of 100 μl solution and replacement by the same volume of 4 M KCl. Any observed change of V_{Cmin} has to be compared with that of unmodified bilayers (Expt. I).

4. Binding of ACTH to Lipid Bilayers at 90 mM Electrolyte Concentration

Increase the ionic strength in the front chamber from 10 to 90 mM by KCl. Stir and record the changing capacitance minimization potential.

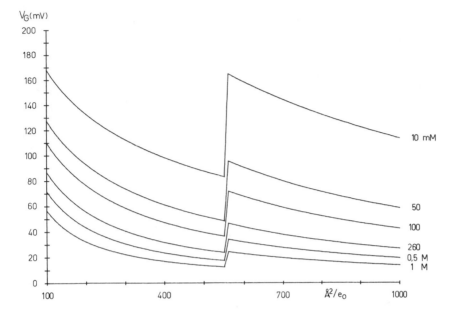

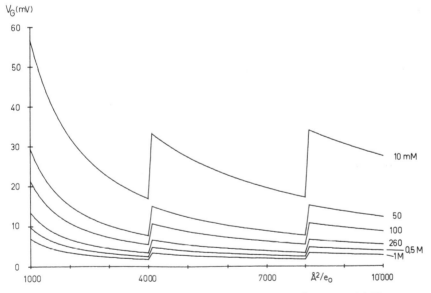

Fig. 6. Theoretical curves representing the Gouy-Chapman surface potential, V_G, as a function of the area/charge for various 1-1-electrolyte concentrations. At the jumps the V_G-axis is expanded by a factor of 2. Surface charge density is found by matching the measured ΔV_{Cmin} values with the difference between the curves corresponding to the initial and final ionic strengths

IV. Discussion

A. Explanation of the Binding Data of $ACTH^{6+}_{1-24}$ to Lipid Bilayer Membranes

In the concentration range used in these experiments ACTH associates with lipid bilayer membranes. This is revealed by the appearance of a positive surface potential on the cis-side, caused by the attachment of positively charged ACTH at the membrane-solution interface. The existence of a "trans effect", i.e., the sensitivity of V_{Cmin} to changes of the ionic strength on the opposite side of the membrane, has to be explained by the appearance of positive charges on the trans-side too. The fact that V_{Cmin} reaches a non-zero equilibrium value, however, shows that ACTH as a whole cannot cross the membrane and equilibrate between the two aqueous phases. These findings indicate that ACTH incorporates into the membranes, with the translocation of some of its charges to the opposite side of the bilayer. As the thermodynamic description of the binding process involves nonlinear, implicit formulas, a quantitative description of the data requires rather extended iterative computer fitting procedures. For a detailed discussion of the numerical results see Schoch et al. (1981).

B. Comments

The ACTH concentration used in this experiment with neutral lipid bilayers is very high compared with physiological values, but it should be noted that binding begins to be measurable at about 10^{-6} M ACTH (Schoch et al. 1981). Biological membranes contain about 10% negatively charged lipids, however, and this causes the surface concentration of sixfold positively charged molecules to be increased by about a factor of 1000 compared to the bulk concentration. That means that with biological membranes the observed incorporation would already occur in the nanomolar region, which is already much nearer to physiological conditions.

If ACTH is really inserted into the lipid core of the membranes this could have two important consequences for the interaction with hormone receptors. First, it is possible that ACTH finds its receptor only after insertion into the lipid portion of the bilayer, by lateral diffusion in the plane of the membrane: then the membrane would really act as an antenna for capturing hormone molecules. A second more exciting and speculative view is the possibility that the actual hormone receptor could be located on the inner surface of the cell membrane or even be a cytoplasmic factor. In these cases the hormonal information would be read off from that portion of the molecule that is translocated through the bilayer. It is known that the message sequence of ACTH is found near the N-terminus. That fits well with our assumption that it is the N-terminal part of ACTH that is probably inserted and made available on the trans-side of the lipid bilayer.

References

Avevard R, Haydon DA (1973) An introduction to the principles of surface chemistry. Cambridge University Press, London New York

McLaughlin S (1977) Electrostatic potentials at membrane-solution interfaces. In: Bronner F, Kleinzeller A (eds) Current topics in membranes and transport. Academic Press, London New York, pp 71–144

Montal M, Mueller P (1972) Formation of bimolecular membranes from lipid monolayers and a study of their electrical properties. Proc Natl Acad Sci USA 69:3561–3566

Schoch P, Sargent DF (1980) Quantitative analysis of the binding of melittin to planar lipid bilayers allowing for the discrete-charge effect. Biochim Biophys Acta 602: 234–247

Schoch P, Sargent DF, Schwyzer R (1979) Capacitance and conductance as tools for the measurement of asymmetric surface potentials and energy barriers of lipid bilayer membranes. J Membrane Biol 46:71–89

Schoch P, Sargent DF, Schwyzer R (1981) The interaction of adrenocorticotropin-(1-24)-tetracosapeptide with black lipid membranes: penetration of the hormone into the membrane (in preparation)

Demonstration of Liposomes by Electron Microscopy

M. MÜLLER

Liposomes are widely used in biology and medicine. Very often, however, the physical characterization of the used vesicles is rather poor, although e.g., size and size-homogeneity are two of the most important parameters of liposomes affecting their behavior in blood circulation and tissue distribution (Juliano and Stamp 1975). By electron microscopy it should be possible to describe the lipid vesicles with respect to size, number of bilayers and changes resulting from lipid/protein interactions down to a resolution of approx. 5 nm. However, in most cases *classical* EM-techniques fail.

Negative Staining is a very rapid and simple technique. Liposomes or membrane vesicles (the size of the objects should not exceed approx. 0.5 μm) are embedded in a matrix of a heavy metal salt. This is achieved by floating a carbon-coated EM-grid on a suitably diluted solution (1 mg/ml in the case of sonicated liposomes). Without removing the excess liquid, the grid is then brought in a short contact with a 2% aqueous solution of uranylacetate or sodium phosphotungstate. The liquid is removed by touching the edge of the grid with a piece of filter paper. The preparation is now ready for the examination in the electron microscope. The appearance of liposomes (without incorporated proteins) in negatively stained preparations depends on the heavy metal salt used (Meister and Müller 1980). With sodium phosphotungstate, lecithine liposomes stick together, are partly flattened and partly fused. Uranylacetate-stained liposomes have an irregular form and are often fused. On the other hand, natural membrane vesicles, which are usually more stable than pure liquid vesicles, may yield better results. Nevertheless, EM-pictures of negatively stained liposomes, reconstituted vesicles or membrane vesicles have to be interpreted with greatest care and are of very little value without a lot of additional information.

Thin Sectioning. By sections through the liposomes or the membrane vesicles, information about the size and the internal structure (number of bilayers) can be expected. Ultrathin sections (less than 60 nm thick) are produced from material embedded in a suitable plastic (usually Araldite/Epon). Fixation of the material is necessary to render it unaffected by the following dehydration and resin embedding procedures. Fixation is usually achieved by incubation in a buffered solution of glutaraldehyde (approx. 3%) followed by osmiumtetroxide (1%). Glutaraldehyde stabilizes the objects by cross-linking the proteins, whereas osmiumtetroxide reacts mainly with the unsaturated fatty acid chains of the membrane lipids. This basic EM-technique, the details of which can be found in every standard EM-textbook (e.g., Meek 1978), yields valuable information on isolated membrane vesicles and also on reconstituted vesicles, if a sufficient amount of proteins are present.

Pure lipid vesicles which are not stabilized by glutarladehyde cross-linked proteins usually do not stand the treatment with OsO_4. In general it is therefore impossible to obtain micrographs of ultrathin sections of liposomes.

Freeze-Fracturing. In frozen bilayer membranes the fracture plane occurs within the hydrophobic part of the membrane. It is therefore possible to obtain information about the distribution and density of intramembraneous particles (mostly proteins) and in some cases to characterize changes due to protein/lipid interactions. Since it is not known how much the vesicles protrude out of the ice, exact size measurements can only be achieved with statistical methods (Van Venetië et al. 1980) assuming spherical shape of the vesicles (which is true up to vesicle diameters of approx. 100 nm).

The main limiting step for the application of the freeze-fracture technique to the characterization of natural and artificial membrane vesicles and liposomes is the cryofixation step. Since liposomes or reconstituted vesicles are usually rather small (25–250 nm), they are easily deformed by ice crystals which are formed during the freezing process. The freezing rates obtained by conventional techniques (by dipping approx. 1 μl of the solution mounted onto a suitable support into melting Freon 22) are not high enough [200 K/s (Van Venrooy et al. 1975)]. The formation of large ice crystals can only be prevented by the addition of cryoprotectors (e.g., 20%–30% glycerol). This may again lead to uncontrolled structural alterations (Zingsheim and Plattner 1976).

Cryofixation. Cryofixation can be considered as a purely physical immobilization procedure if it brings about solidification (of water or solutes) in a microcrystalline state in the absence of any cryoprotectives. This goal is achieved if freezing rates of approx. 10,000 K/s are obtained. Under such conditions, structural alterations during the fixation procedure are minimized. However, it remains an open question whether the molecular reorganization of membranes is prevented.

In practice such high freezing rates are realized by the spray-freezing technique (Bachmann and Schmitt 1971) and by "propane jet-freezing" (Müller et al. 1980). The equipment for both techniques is commercially available (Balzers, Fürstentum Liechtenstein).

Spray freezing uses a spray gun to transform the vesicle solution into very small droplets ($\sim$ 5–20 μm). These droplets are directed into liquid propane at 88 K. Because of their optimal shape (Moor 1965), a very high freezing rate is obtained within the smaller droplets (5–10 μm). The frozen droplets are glued together with ethyl benzene at 188 K, transferred to suitable supports and stored in liquid nitrogen for further use.

For propane jet freezing, a very small amount (1 μl) of the vesicle solution ($\sim$ 2–10 mg lipid/ml) is sandwiched between two thin copper plates (0.08 mm). This sandwich is placed between two nozzles through which a fast stream of cold (88 K) propane is blown onto the surface of the sandwich. With this technique, very high freezing rates are obtained within the entire sample.

Cryofixed specimens can be analyzed by freeze-fracturing following a purely physical protocol, or by freeze-substitution. Freeze-substitution offers a way to prepare the samples for thin sectioning. However, the problems of chemical fixation are avoided almost completely.

Freeze-Fracturing is performed according to standard procedures, adapted to the available equipment (Moor 1965).

Freeze-Substitution has become an important preparation technique only since the advent of the rapid freezing techniques mentioned above. The propane-jet frozen sandwiches are better suited for the treatment by freeze-substitution because the freezing quality is more uniform than in the spray-frozen droplets, where only the smaller ones are sufficiently well frozen.

Fig. 1. Sonicated EYL (Egg Yolk Lecithine)-Liposomes (2 mg/ml), frozen in a propane-jet and freeze-fractured. Freeze fracturing yields information on the shape of the vesicles and allows an estimation of the size by statistical methods (Van Venetië et al. 1980)

Fig. 2. The same preparation treated by freeze-substitution. It shows that the vesicles are monolamellar. The size distribution of thinsectioned vesicles is easily obtained using the method of Rose (1980)

Fig. 3. EYL-Liposomes prepared by controlled dialysis (Milsmann et al. 1978) of the detergent (cholate) in the presence of an iron dextrane complex. The vesicles are propane-jet frozen and freeze-substituted

Fig. 4. Liposomes prepared with the controlled dialysis procedure, the detergent being Oxtylglycosid. The vesicles are much larger than those in Fig. 3 and exhibit a more irregular structure

Fig. 5, 6. Dimeristoyl-lecithine vesicles, prepared by detergent removal below the transition temperature. Characteristic structures are recognized in the freeze-fractured (Fig. 5) and the freeze-substituted (Fig. 6) samples

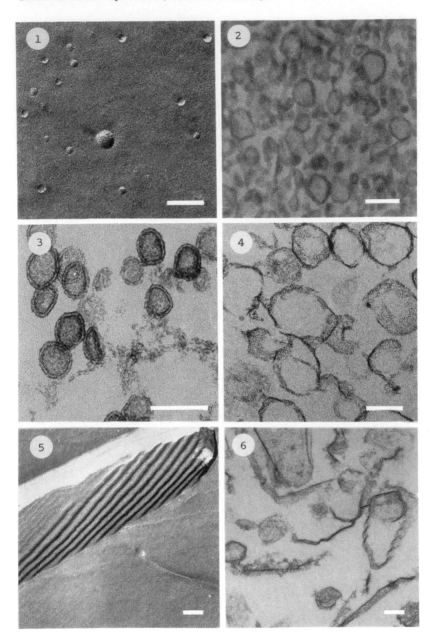

There is no commercial freeze-substitution equipment available. It is, however, easily improvised. Satisfactory results are obtained by the following procedure (Müller et al. 1980): The frozen samples are immersed into 1 ml of the substitution medium at 178 K. The substituent consists of methanol containing 0.5% uranylacetate, 1% OsO_4 and 3% glutaraldehyde. After 8 h the temperature is raised to 210 K and finally to 243 K for 8 h. At this temperature the substituent is replaced by acetone. Liposomes are easily pelleted with a Beckmann Microfuge B operating in a deep freezer at 243 K. The samples are then placed in a refrigerator at 280 K where they are embedded in Araldite/Epon. Ultrathin sections are stained with UO_2AC and/or lead citrate (Meek 1978).

The scope of applications of the electronmicroscopical techniques based on cryofixation (freeze-fracturing and freeze-substitution) is illustrated by Figs. 1–6. (In all figures the bar represents 100 nm.)

The results shown indicate that on the basis of a purely physical cryofixation (without chemical fixatives and cryoprotectants) the structure of liposomes can be described.

References

Bachmann L, Schmitt WW (1971) Proc Natl Acad Sci USA 68:2149

Juliano RL, Stamp D (1975) Biochem Biophys Res Commun 63:651–658

Meek GA (1978) Practical electron microscopy for biologists, 2nd ed. John Wiley & Sons, New York

Meister N, Müller M (1980) Proc 7th Europ Congr Electron Microscopy. The Hague 2: 642–643

Milsmann MHW, Schwendener RA, Weder H-G (1978) Biochim Biophys Acta 512: 147–155

Moor H (1965) Balzers Report BB 800 006 DE

Müller M, Marti Th, Kriz S (1980) Proc 7th Europ Congr Electron Microscopy. The Hague 2:642–643

Müller M, Meister N, Moor H (1980) Mikroskopie (Wien) 36:129–140

Riehle U (1968) ETH Diss Nr. 4271, Zürich

Rose PE (1980) J Microsc 118:135–141

Van Venetië R, Leunissen-Bijvelt J, Verkleij AJ, Ververgaert PHJTh (1980) J Microsc 118:401–408

Van Venrooy GEPM et al (1975) Cryobiologie 12:46

Zingsheim HP, Plattner H (1976) In: Korn ED (ed) Methods in membrane biology, vol 7. Plenum Publ Co, New York

Molecular Biology
Biochemistry and Biophysics

Editors: A. Kleinzeller, G. F. Springer, H. G. Wittmann

Volume 25
Advanced Methods in Protein
Sequence Determination

Editor: S. B. Needleman
With contributions by numerous experts.

1977. 97 figures, 25 tables. XII, 189 pages
ISBN 3-540-08368-5

The determination of protein sequences has become so
commonplace that, as more laboratories have entered this
area of study, the sophistication of the technology has,
in fact, progressed to make use of physical properties not
previously utilized for this purpose. Earlier manual techniques have become automated; current instrumentation
operates at higher parameters and with greater precision
than before. Thus the present volume supplements the
earlier one in presenting details of the more advanced
technologies (optical, high pressure, X-ray, immunology
etc.) being used in sequence determination today.

Volume 29: E. Heinz
Mechanics and Energetics
of Biological Transport

1978. 35 figures, 3 tables. XV, 159 pages
ISBN 3-540-08905-5

This book presents the interrellations of mechanistic
models on the one hand and the kinetic and energetic
behavior of transport and permeatin processes on the
other, using the principles of irreversible thermodynamics. The advantages of each method are compared. The
special aim is to show how to appropriate formulas can be
transformed into each other, in order to recognize in what
way the kinetic parameters correspond to those of
irreversible thermodynamics.

Springer-Verlag
Berlin
Heidelberg
New York

Membrane Biochemistry

A Laboratory Manual on Transport and Bioenergetics

Editors: E. Carafoli, G. Semenza
With contributions by numerous experts

1979. 45 figures, 6 tables. X, 175 pages
ISBN 3-540-09844-5

Contents: Nonelectrolyte Transport in Small Intestinal Membrane Vesicles. The Application of Filtration for Transport and Binding Studies. – Transport of Sugars in Bacteria. – Net Na^+ and K^+ Movements in Human Red Blood Cells After Cold Storage. – Calcium Transport in Resealed Erythrocytes and the Use of a Calcium-Sensitive Electrode. – Calcium Transport in Sarcoplasmic Reticulum Vesicles Isolated from Rabbit Skeletal Muscle. – Preparation and Assay of Animal Mitochondria and Submitochondrial Vesicles. – Measurement of Cytochrome Kinetics in Rat Liver Mitochondria by Stopped Flow and Dual-Wavelength Spectrophotometry. – Proton Translocation Catalyzed by Mitochondrial Cytochrome Oxidase. – Determination of the Membrane Potential and pH Difference Across the Inner Mitochondrial Membrane. – Anion Transport in Mitochondria. – Calcium Transport in Mitochondria. – Redox Intermediates Between O_2 and H_2O. – Photophosphorylation with Chromatophore Membranes from *Rhodospirillum rubrum*. – Oxygen Evolution and Uptake as a Measure of the Light. – Induced Electron Transport in Spinach Chloroplasts. – The Function of the Purple Membrane in *Halobacterium halobium*. – Characterization of Ionophores Using Artificial Lipid Membranes. – Characterization of Neutral and Charged Ionophores Using Vescicular Artificial Liquid Membranes (Liposomes).

The editors of this book, themselves well-known experts in the field, have assembled a broad spectrum of experiments in membrane transport and membrane bioenergetics. Basic techniques, as well as new and more sophisticated procedures are described in detail, with special attention paid to those which can be applied to a wide range of projects. The methods presented enable even readers with little knowledge of membrane biology to carry out correctly the most important experiments in the field.

The experiments reported have all been successfully tested in the Advanced Courses 29, 41, 45 and 52 of the Federation of European Biochemical Societies for Membrane Biochemistry, Transport and Bioenergetics, organized by the editors at the Swiss Institute of Technology, Zurich, between 1975 and 1978.

Springer-Verlag
Berlin
Heidelberg
New York